Molecules of Life & Mutations
Understanding Diseases by Understanding Proteins

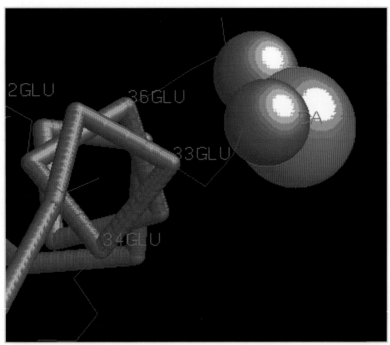

One Calcium ion specifically bound to a Ca^{2+} pocket in the adhesion protein **E-Selectin** (PDB 1ESL)

Siegfried Schwarz

Molecules of Life & Mutations

Understanding Diseases by
Understanding Proteins

475 figures in color, 2002

KARGER

Professor Siegfried Schwarz, M.D.
Institute of Pathophysiology
University of Innsbruck, School of Medicine
Fritz-Pregl-Straße 3
A 6020 Innsbruck · Austria
siegfried.schwarz@uibk.ac.at

The **Brookhaven PDB** was accessed over the Internet, and proteins that appeared important and instructive were selected for inclusion in this book. Their coordinate files (marked herein with the respective PDB accession number) were loaded and, with **RasMol**® software, visualized in such a fashion that the point to be taught, the didactic 'message', and detail are highlighted. I deliberately restricted myself to two colors and their gradations of shade in order to focus the reader's eyes on the essential elements of a given molecule. Using the accession code, the reader can find the names of those who elucidated each protein's crystal structure and created the corresponding coordinate file. **I would like to give here all due credit and thanks to these authors for their immense work in elucidating these structures for us.** In addition, details, background information and primary literature on the proteins referred to can be easily called up on the Internet. The address of the Brookhaven PDB's 'Advanced Search 3D-Browser' is http://www.rcsb.org/pdb/cgi/queryForm.cgi. Pictures with smoothed molecular surfaces were created using the software **Chime**® by **MDL** Informations Systems Inc.

All rights reserved. No part of this publication may be translated into other languages, reproduced or utilized in any form or by any means, electronic or mechanical, including photocopying, recording, microcopying, or by any information storage and retrieval system, without permission in writing from the publisher.

Printing of this book was kindly supported by  **and** *serono* International SA.

Copyright ©2002 by S. Karger AG, P.O. Box, CH-4009 Basel (Switzerland)
Printed in Austria by Athesia-Tyrolia Druck, Innsbruck
ISSN 0074-1132
ISBN 3-8055-7395-2

Acknowledgements

The author would like to express here his gratitude to the University of Innsbruck, particularly to Professor Georg Wick, head of the Institute of Pathophysiology, for his continuous support, as well as to all the colleagues at the University Computer Center who take care of a continuous and reliable Internet access. Furthermore, I would like to thank my technician, Irene Gaggl, who spent many hours at the computer in downloading PDB files for me.

My deep gratitude should also be extended to Ms. Rajam Csordas-Iyer, Innsbruck, who prepared a first English draft of my German version of this book, and to Ms. Kat Occhipinti, La Jolla/San Diego, who did the final editing and polishing of the manuscript. Both persons, compelled by the subject, carried out these efforts with great enthusiasm and dedication.

Also, I want to gratefully mention Dr. David Rodbard, NIH, Bethesda, Md., USA, who taught me in his Laboratory of Theoretical and Physical Biology the importance of 'crystal-clear thinking'.

Last but not least, I want to thank three friends in Innsbruck: Architect Joachim Fanta †, for providing me the 'flower and bees' drawing, Dr. Eugen Preuss and Dr. Heidrun Recheis for having helped me in many situations of sheer technical difficulty and despair with computer hard- and software. It is too sad that Joachim can not see this book anymore. May his widow Eva see it and think of him.

Kat Occhipinti, who has seen only my text but so far none of the pictures, wrote me, after completion of her work, the following email:

Dear Siegfried!
This book takes an approach to science that is basic and beautiful and, for me, a reflection of the majesty of God who created, and teaches us with, a world of unending mystery and beauty both galactic and microscopic. Honor to you for showing us the beauty of that world! It was such a pleasure for me to work on the text of this book!

Thank you so much, Kat, this was so kind and insightful!
Siegfried

Dedications

Beauty

Dedicated to my beloved wife Rita and my wonderful daughters Michaela, Christina and Marielies

Form and Function

Dedicated also to my parents, particularly to my father who died far too early, but who, as a teacher, had taught me to teach

May this book help to educate doctors to serve their patients in a throughtful and lucid way!

In this sense, I want to dedicate this book also to my cousin Christine who became severely handicapped life-long, and to my brother-in-law Ludwig who died unneccessarily early, both, because their diseases were not recognized in time.

Introduction

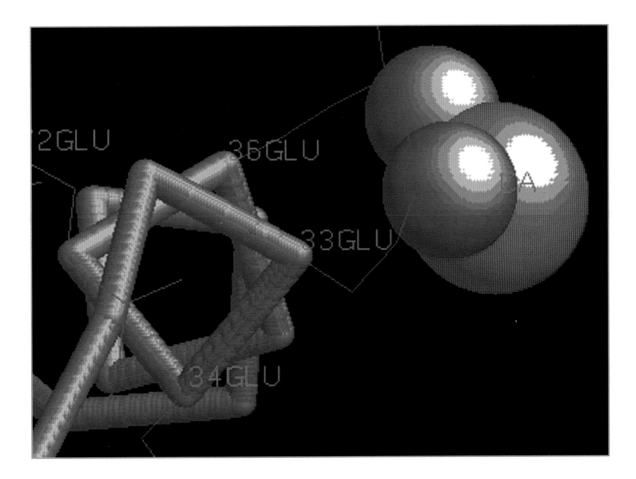

Molecules of Life and Mutations has been written with the intention of providing insight into biomolecules, their materiality, their different sizes and complexities in shape and structure. The book is comprised of images that should make material the magnitude of the microscopic and teach us perhaps more than any other medium about the mechanisms and components of cellular systems that have been until now to most people theoretical. Pictures dynamically present the simplicity as well as the complexity of the shape, structure and function of the molecules of life as well as prove their interdependence. Seeing the concrete and 'real' structure of a molecule should explain to the reader its function in a cell or in an organism. It should become clear that no biomolecule by itself is effective, but becomes so only when it enters an interactive relationship with another specific molecule. This interaction requires a structural specificity, i.e. portions of the two molecules involved that match each other in form. The following pictures will show that these portions are not based on rough complementarity, but virtually on atomic precision.

For the most part, such interactions should not be long-lasting, but rather short and transient, i.e. it must be possible to discontinue an interaction after a short period of time. Noncovalent binding offers the best condition.

Random micro- and macromovements as well as vectorial transport processes of a molecule in the various compartments of a cell or an organism provide the opportunity for interaction. In this process, the probability of collision is determined by the concentration of one or both of the molecules that is ready to react. Noncovalent binding results from a collision of the two when they match in form at certain points. The basic pattern consists primarily of a small molecule (a ligand), which is not necessarily a protein, and a larger one, which is always a protein, that becomes activated by the former (also called an agonist). The latter can, by virtue of the inherent plasticity of most proteins, undergo a conformational change through binding of the agonist. The agonist activates (makes competent) the protein by inducing a specific conformational change that enables it to carry out its function, i.e. to interact with a further molecule, for example a membrane receptor with a signal transduction protein, or a transcription factor with a specific target gene.

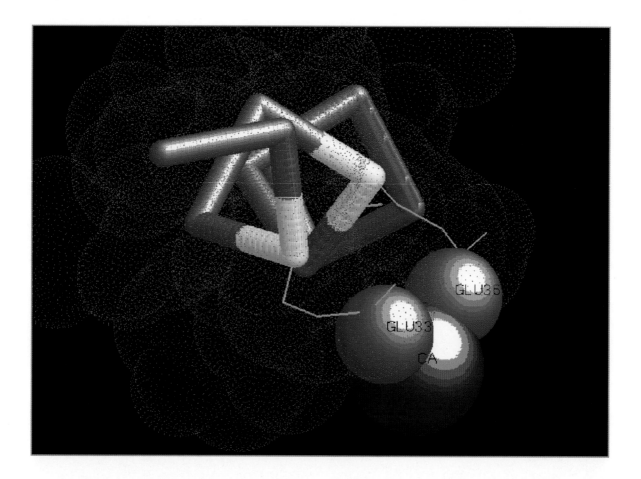

The effect of an activated protein ceases on its own eventually because the condition of being functionally competent is, in itself, unstable due to the dissociation of the allosterically activating agonist from the activated target protein, which thus reverts once again to its initial inactive or resting conformation. This switching back and forth between two different conformational states is a characteristic property of certain proteins (switch proteins), but not a feature of the nonprotein ligands. Of course, these two states are only the extremes of a whole spectrum of many intermediate conformations. Nonprotein ligands have a rigid structure, as best demonstrated by the example of the smallest ligand, a single calcium atom, which will be detailed in the chapter 'Small Molecules'. It should be pointed out that peptides and proteins can also be ligands, and that ligands can induce both activation and inactivation of proteins.

This sequence of events and the temporal limitation of molecular interactions and their effects is the fundamental principle of cellular signaling, the communication between cells of an organism, and also of the cellular reception of signals from the outer world, such as light, nutrients (xenobiotica), toxins, microbes, or drugs, among others. Proteins are the most important mediators of signals. Specialized signal-receiving proteins are called receptors. Proteins are encoded by genes (DNA) and, in turn, as so-called transcription factors, they regulate gene expression. The importance of signaling is demonstrated by the fact that about 30% of the nearly 30,000 human genes code for signaling proteins. A multi-cellular organism is not viable and is inconceivable without communication, or signaling between its constituent cells.

Mutations in genes lead to diseases, often severe and even fatal, because mutated genes express functionally altered proteins. This is true of genes in general, and of genes encoding signaling proteins in particular. The exact nature of the functional disturbance cannot be read from the gene, but only from the structure of the protein involved, if this is known. Until a few years ago, however, the structure of most proteins was unknown.

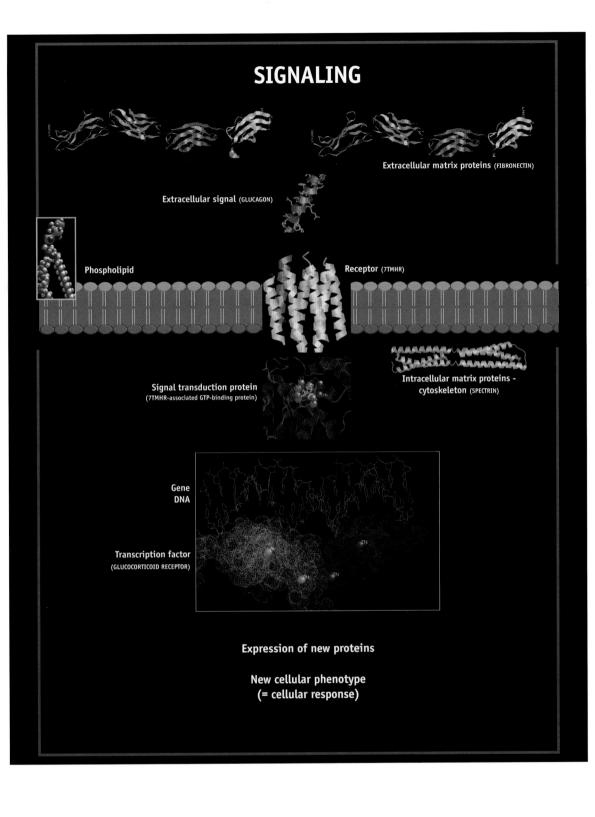

Advances in molecular biology have permitted the production of large amounts of a specific protein (in recombinant form) and thus the production of protein crystals. X-ray diffraction analysis has made it possible to reconstruct a protein's structure with atomic detail in a three-dimensional form. The resolution of these techniques is below 2 Å, i.e. the same as the van der Waals' radius of sulfur, carbon, nitrogen and oxygen atoms. In the Brookhaven Protein Data Bank (PDB), the individual X, Y and Z coordinates of hydrogen, carbon, oxygen, nitrogen and all the other constituent atoms of approximately 10,000 proteins are so far deposited. These data have become generally accessible through the Internet, and can now, with the help of appropriate molecular modeling software, be manipulated, rotated, zoomed and visualized (in stereo!) on a personal computer.

If one knows the domain of a protein responsible for each function (e.g. the substrate-binding domain of an enzyme, the ligand-binding domain of a receptor, the DNA-binding domain of a transcription factor), or which of the multiple domains of a protein is afflicted by an abnormal conformation, it should be possible to directly deduce the biological consequence of the altered conformation. On one hand, mutations are productive and important for successful evolution and the creation of newer forms of life in the course of millions of years; yet, conversely, mutations can also be counter-productive in that they can cause diseases, and thus are of medical relevance.

The size of molecules covers a vast range, encompassing at one end of the spectrum tiny calcium atoms (with a molecular mass of 40), and at the other end large lipoproteins or chaperones (with a molecular mass above 1,000,000). It will become clear that where a protein is not large enough to form a specific structure, means will be employed, such as linking together smaller protein modules (domains) and even di- and multimerization of whole proteins. It will also become clear that the sizes of interacting molecules or portions thereof should be commensurate: A small ligand (e.g. a calcium atom or a steroid hormone) fits into the small ligand-binding pocket of a protein, while a larger ligand (e.g. a protein hormone, such as the human chorionic gonadotropin, hCG) fits only in a correspondingly larger groove, which can be formed only by a multimodular protein.

It has already been mentioned that fitting biomolecules requires atomic precision. Thus, it should be clear that even the smallest mutation, a so-called point mutation or change of a single base in a gene, can disturb this precision at the protein level. My primary concern was to make this obvious to the reader, in particular physicians, medical students and biologists. Moreover, pharmaceutical chemists can, through their knowledge of the 3-dimensional structure of a pocket in a macromolecule, construct artificial ligands (rational drug design), quasi-negative copies, with the ability to occupy this site and thus function as agonists or antagonists. Perhaps above all, every reader regardless of background and specific interest, may be captured and fascinated by the beauty of the structure and architecture of biomolecules, particularly proteins.

Contents

Ligands – Small Molecules	1
Nonpeptidic 'Small Molecules'	1
Calcium	1
Catecholamines	5
Morphine	6
Adenosine Phosphates	7
Steroids	8
Estrogens	8
Progesterone	9
Thyroid Hormones	10
Enzyme Inhibitors	12
DNA-Binding Substances	13
Carcinogens	13
Cytostatic Agents	14
Ionophores	15
Peptidic and Proteohormones	17
Peptidic Toxins	27
Pictorial Representations of Proteins (Insert)	i
Ligand	ii
Transforming Growth Factor-β	ii
Ligand-Binding Protein	xvii
Human Chorionic Gonadotropin Receptor	xvii
Ligand-Binding Proteins – Large Molecules	31
Enzymes	31
Membrane Receptors	43
Signal Transduction Proteins	53
Multimodular Adhesion Proteins	61
DNA-Binding Proteins	65
DNA/RNA-Processing Enzymes	68
Antibodies against DNA	70
DNA-Bending Proteins (HMG-Box Proteins)	70
Transcription Factors	71
Tumor Suppressor Proteins	80
Telomere-Binding Proteins	82
Homeobox Proteins	83
Immunological Proteins	87
Antibodies and B-Cell Receptors	88
MHC Proteins	92
T-Cell Receptor	94
Immunophilins	96
Miscellaneous Macromolecules	97
Apolipoproteins	97
Crystallins	98
Collagens	99
Hemoglobin, Spectrin	100
Stress Proteins, Chaperones and Heat Shock Proteins	101
Proteasomal Proteins	102
Viral Coat Proteins	103
Epilogue	105
Index	107
Internet Resources	111

Ligands – Small Molecules
Nonpeptidic 'Small Molecules'
Calcium

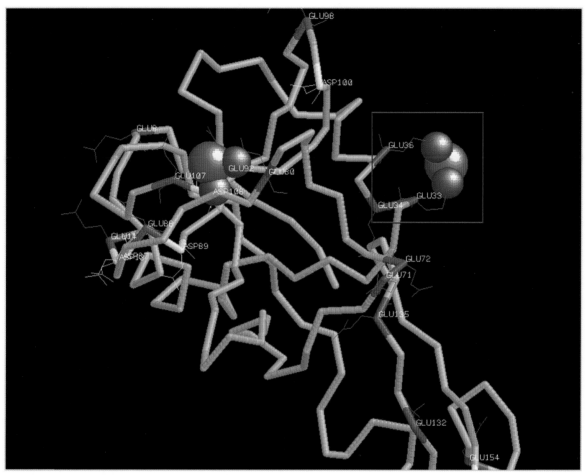

Two Calcium ions each specifically bound to a Ca^{2+} *pocket* in the adhesion protein **E-selectin**

(PDB 1ESL)

Calcium is the smallest of all ligands: A single atom or, more specifically, a positively charged bivalent ion.

Illustrations on the previous and following pages (1–4) show a single calcium atom bound to large calcium-binding proteins, such as E-selectin or calmodulin. Ca^{2+} is represented here as a space-filling atomic model (referred to as CPK in this book), in contrast to the protein of which only the tube-like alpha-carbon backbone with the amino-acid (AA) side chains in *wireframe* representation is shown. As can be seen, acidic AAs (Glu = glutamate and Asp = aspartate) are positioned in the overall structure of the protein in such a way that their carboxylate ions (oxygen atoms, here depicted as grey CPKs) come so close to each other that the space between them can be filled precisely by nothing but a calcium ion.

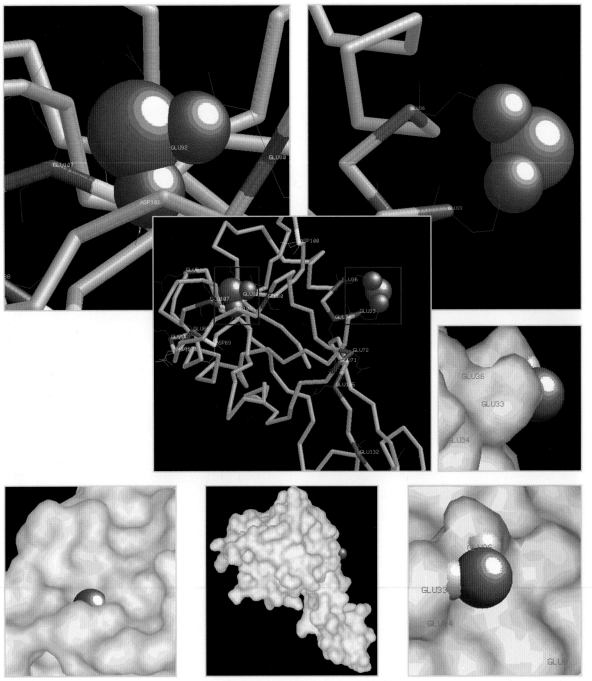

Calcium ion bound to a Ca^{2+} pocket in the adhesion protein E-selectin

(PDB 1ESL)

There must be between two and six such carboxylate ions to form a high-affinity calcium-binding pocket. Since a carboxylate ion is negatively charged, while the calcium ion is positive, an electrostatic attraction between the two can be established. As a consequence of the binding of one Ca^{2+} ion, the carboxylate ions involved are shifted slightly away from their original position in the protein, resulting in a slight displacement of the corresponding amino acids, which causes (partly via the backbone) a conformational change of the entire protein, even in domains distant from the calcium pocket. This latter effect provides the basis for the protein to interact with a further target molecule, e.g., with myosin light-chain kinase (MLCK) in the case of Ca^{2+}:calmodulin (see pictures in chapter 'Signal Transduction Proteins').

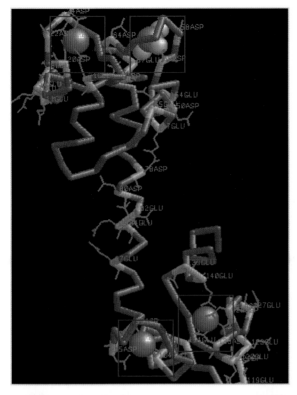
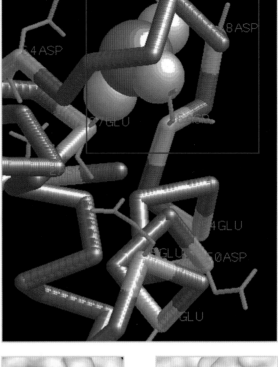
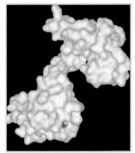
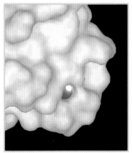
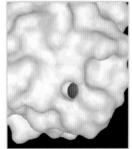
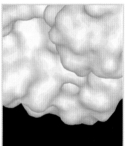

Calcium ions, each specifically bound to a Ca^{2+} *pocket* in the signal transduction protein **calmodulin**

(PDB 1ESL)

The pictures on pages 1 and 4 show how in E-selectin, a calcium ion is held by two carboxylate ions of the glutamate residues 33 and 36 which are correctly positioned one above the other. They are so positioned because Glu33 and Glu36 are parts of an alpha helix of the protein, Glu36 being exactly one turn of the pitch below Glu33 in the helix. The other calcium pocket of E-selectin, formed by Glu80 and Asp106, can be seen on page 2.

Further, the pictures show that a protein can possess several calcium pockets: E-selectin has two such pockets, while calmodulin, pictured above, has four. There are cooperative allosteric effects between the individual pockets, i.e. when one is liganded with Ca^{2+}, the affinity of the next for Ca^{2+} may increase. Thus, the entire calmodulin molecule successively undergoes discrete conformational changes.

E-selectin is a membrane protein projecting into the extracellular space and thus displaying cell adhesion function. Calmodulin is an intracellular cytosolic, soluble signal transduction protein. In addition to E-selectin and calmodulin, there is a vast number of other Ca^{2+}-sensitive or Ca^{2+}-dependent proteins.

The concentration of free calcium in blood is measured via a calcium-sensing receptor in the membrane of parathyroid cells. An inactivating mutation in the gene for this protein leads to hyperparathyroidism with severe bone resorption as a consequence. Ca^{2+}-dependent proteases can be activated by excessively high concentrations of cytosolic Ca^{2+}, which leads to cell breakdown, the basis of the neurotoxicity of specific neuroexcitatory substances.

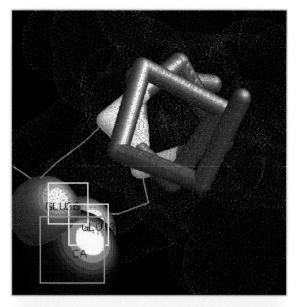
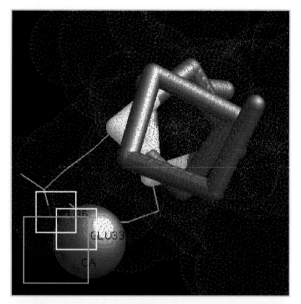
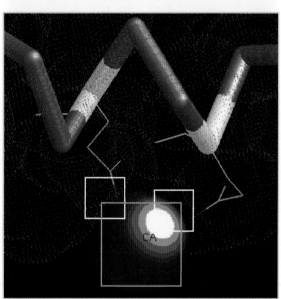
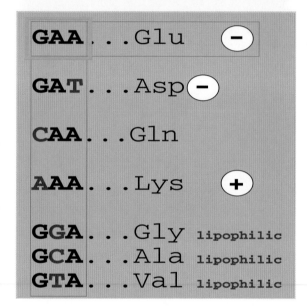

If only one of the carboxylate ions in a calcium pocket is shifted (e.g. by a more distant mutation in the protein) or missing (e.g. by a mutation in the pocket itself, which converts for example an Asp = aspartate to an Asn = asparagine, or Glu = glutamate to a Gln = glutamine), the Ca^{2+} binding ability is abrogated and thus the activatability of the entire Ca^{2+}-dependent protein. Clearly, **atomic precision** is an indispensable prerequisite for the proper functioning of ligand-binding/switch proteins.

The picture on the bottom right shows examples of how point mutations in the triplet **GAA**, which encodes glutamate (Glu), can result in (1) an amino acid such as Asp (aspartate) which, although still having an acidic carboxylate group, is one C-atom shorter than Glu; (2) a lipophilic amino acid such as glycine (Gly), alanine (Ala) or valine (Val) that is unable to bind Ca^{2+}, or (3) a positively charged amino acid such as lysine (Lys) that would even reject Ca^{2+}.

Catecholamines

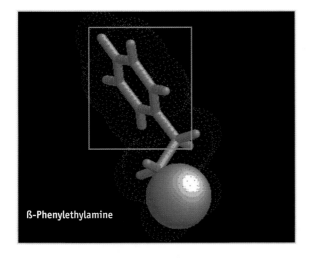
ß-Phenylethylamine

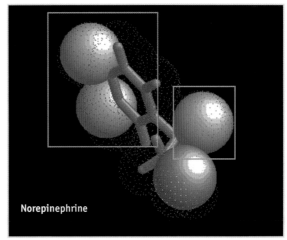
Norepinephrine

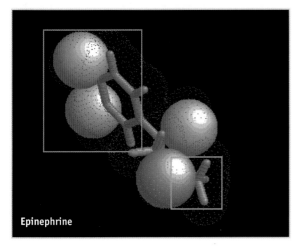
Epinephrine

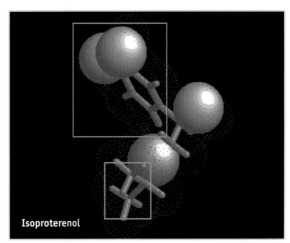
Isoproterenol

Catecholamines are important *hormones* and *neurotransmitters*. They are hydrophilic molecules and react on one hand with adreno(re)ceptors, and on the other with amine uptakers and various enzymes that synthesize or metabolize them [for example, convert norepinephrine to epinephrine, or degrade norepinephrine to normetanephrine, i.e. the monoamine oxidase (MAO)].

The catecholamine analogs ordered here from top left to bottom right, exhibit an increasing structural complexity, both with respect to ring-hydroxylation (O:light magenta, CPK) and to methyl substitution of the amino group's nitrogen (N:grey, CPK). Without ring hydroxylation (e.g. in β-phenylethylamine), the affinity for adrenoeptors is extremely low. With the ring hydroxylated, however, its affinity is high, yet unselective between α and β adrenoceptors. Increasing methyl substitution in the amino group (the difference between epinephrine and norepinephrine!) lessens the affinity for the former and simultaneously increases it for the latter, thus making the substances β-selective. This is desirable and necessary, since α and β adrenoreceptors are expressed in different cell types and therefore mediate completely different effects. These examples demonstrate once more that apparently minor atomic changes can have major biological and pharmacological consequences. Indeed, even if, in a given molecule, the bond of a certain substituent is turned into the opposite direction, the receptor binding ability and thus the function of the molecule can be dramatically changed, a phenomenon called *stereospecificity* or *stereoselectivity!*

An inactivating mutation in certain adrenoceptor genes can lead to hereditary obesity, another to hypertension. The so-called β-blockers are among the most widely used drugs given to treat patients with high blood pressure.

Morphine

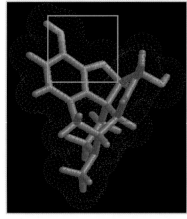 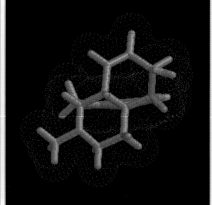 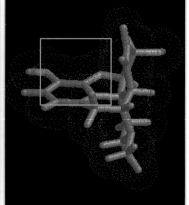

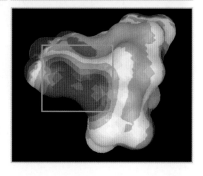

Morphine in different views; the right lower picture (180° flipped in the x-axis relative to that above) shows the electrostatic potential, red being electronegative; note the coincidence with the two oxygen atoms (in frames)!

In contrast to proteins, many of which have a certain *plasticity,* small molecules have a *rigid* structure. Often they are bent more or less strongly at one point (as can also be seen in catecholamines), sometimes they consist of two or more parts that are, to a certain extent, rotatable relative to each other or fixed in a certain angle. Such a rigid chair-like structure is typical for *alkaloids,* i.e. substances that can only be synthesized by plants, such as morphine, here viewed from three different perspectives.

Morphine is a substance that, like catecholamines, interacts with 7 transmembrane helix receptors (7TMHRs, see chapter on 'Membrane Receptors'). As can be seen from the picture above, morphine is about twice as large as epinephrine, although the receptor is of equal size and quite similar in structure to adrenoceptors. Opiate receptors also occur as several iso-receptors (or subtypes): δ, μ, and κ opiate receptors. The natural ligands, however, are *endogenous opiates* and not exogenous morphine, and are *peptides!* Obviously, peptides can adopt a morphine-like conformation (or vice versa). Otherwise, they would have to react with quite different amino acid residues in the opiate receptor. Nonetheless, they lead to the same *productive conformational* change and thus to the same cellular response.

> Opiate receptors are eminently important for the suppression of transmission in pain sensory neurons. Opiate receptors in other neurons of the central nervous system also mediate euphoria or depress respiration. Morphine and its derivatives have been known to mankind for thousands of years as agents that induce euphoria and/or suppress pain.

> Morphine is regarded as an *opportunistic ligand* of opiate receptors, i.e. as a substance that, by chance, fits into a protein that has actually been designed to recognize another ligand, namely an endogenous peptide. Toxins and viruses are also opportunistic ligands.

> The former *misuse* a specific membrane receptor of a specific cell type by inactivating it, the latter in that they find entry (via *receptor-mediated endocytosis*) into the cell (= *infection*).

Adenosine Phosphates

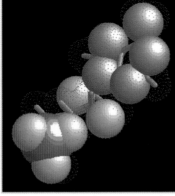

AMP cAMP ATP

Adenosine phosphates are about the same size as catecholamines and have a similar general structure, yet of different atomic composition in that they possess 5 nitrogen atoms as well as 1 (AMP, cAMP), 2 (ADP) or 3 (ATP) phosphate residues. A phosphate renders this molecule negatively charged (in contrast to the positively charged amino group in catecholamines).

Adenosine phosphates react with different proteins: cAMP (but not AMP, ADP, ATP) reacts with the regulatory subunit of protein kinase A in the sense of an allosteric activation that leads to dissociation of the catalytic subunit; ATP (but not AMP, ADP) reacts as substrate with adenylate cyclase; ATP (but not AMP and ADP) reacts with ATPases (e.g. membrane pumps) and all kinases in the sense of an energy-rich phosphate donor. In ATPases, the hydrolysis of ATP to ADP generates the energy required to induce the conformational change necessary to translocate an ion over the barrier of a biological membrane. In kinases, the liberated phosphate residue can be transferred to the substrate to be phosphorylated. The remaining ADP is no longer bound tightly enough and therefore dissociates from the ATPase or the kinase.

Proteins with adenosine pockets can thus distinguish with great selectivity between AMP, cAMP, ADP, and ATP.

Cyclic AMP is among the first of the so-called second messengers that have been identified. Ca^{2+} ions, described above, also belong to this class of molecules, among a few others. Second messengers convey the message of the primary extracellular signal to the interior of a cell if it is hydrophilic, and thus cell membrane-impermeable. Second messengers arise on the inner side of the membrane, immediately below or downstream of a membrane receptor.

From the structural analogs depicted here, it can be appreciated once more that apparently minor molecular differences have far-reaching biological consequences, because even minimal structural differences (remember, stereospecificity) confer binding specificity on such substances. The difference between AMP and cAMP is hardly noticeable at a first glance, yet this molecular difference is perceived very precisely by the corresponding proteins.

If too much cAMP is produced in a cell, it leads to severe diseases, such as cholera or hypertrophy of an organ (thyroid gland, among others) depending on the cell type affected.

Steroids
Estrogens

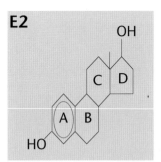

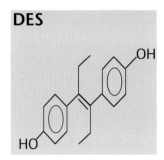

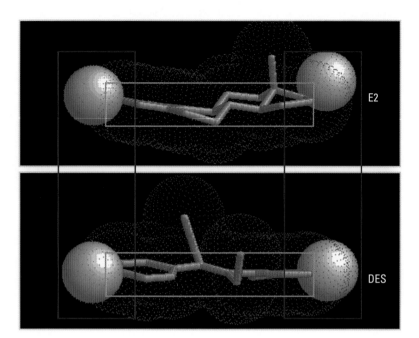

Estradiol-17ß (E2) and diethylstilbestrol (DES)

Small molecules always have a rather rigid structure such that some may consist of portions that are more or less kinked towards each other, as already mentioned (catecholamines, morphine, adenosine phosphates). Often, however, they are flat as a board. The latter construction principle is typical for steroids and DNA-intercalating substances, to mention a few.

Two small ligands can have similar or even the same binding properties and thus the same effectiveness, even when the details are quite different relative to structure and atomic composition. Attention has already been drawn to this important fact in the case of morphine and endogenous opiate peptides. Here, this principle will be explained for a steroid hormone, namely the female sexual hormone estradiol-17b (E2). The basis for this phenomenon is that it is not the ligand as a whole molecule with all its atoms that reacts with a protein in question (e.g. the *estrogen receptor*), but only a few selected atoms of the ligand (= reactive groups = the *pharmacophores*, e.g. the oxygen atoms in E2: magenta CPK).

Through appropriate synthetic steps, it is possible to arrange these as reactive-identified atoms at the proper orientation to each other on a non-steroidal scaffold, e.g. a *stilben* backbone, and reproduce the function of E2 (*molecular mimicry*). Thus, the diethylstilbestrol (DES) pictured here is also an estrogen, but a nonsteroidal estrogen (estrogen analogs in the proper sense always have a steroidal structure!). A number of estrogen-active nonsteroidal substances can be found in many plants, in soya, among others, and are gaining increasing interest in estrogen replacement therapy schemes (ERT) for postmenopausal women.

Note that in DES, rings B and C are open and ring D is an aromatic 6-ring, whereas rings B and C in E2 are closed and ring D is an unsaturated 5-ring. Moreover, the second hydroxyl group of DES is in position 16b, not 17b. By a slight torsion of the ring levels of A and E, the second hydroxyl group comes obviously (and with a certain amount of tolerance) to the correct (i.e. as in E2) position. It becomes clear that the usual 2D chemical formulas do not convey the right impression about structure-function relationships. For that, one clearly needs 3D representations, as shown here.

Progesterone

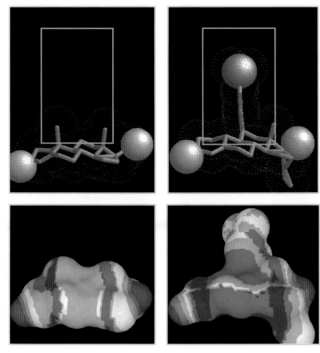

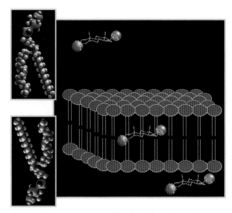

Above: **Progesterone** 'diffusing' through a phospholipid bilayer cellular membrane (insets show two phospholipid groups enlarged)

Progesterone   RU486

Left lower pictures show the electrostatic potential surfaces of progesterone and RU486. Note the similarities in position and relative distance of the 2 oxygen atoms (red)!

Progesterone is another major member of the 'family' of steroid hormones. It binds to the progestin *receptor* and also to *corticosterone-binding globulin* (CBG), as well as to a number of *steroid metabolizing enzymes*, such as those that convert it to 17a-hydroxyprogesterone or degrade it to pregnanediol.

In small molecules, substituents may play a much more decisive role in structure and function than occasionally a similar alteration in a large protein. For example, a small substituent such as the one seen in RU486 pointing upwards (with all other atoms remaining almost unchanged) can convert the agonist progesterone into a receptor antagonist. Such a modification can, in addition, also change the specificity: RU486 can, as an antagonist, also occupy the glucocorticoid receptor, whereas progesterone is selective for the progestin receptor.

The above-mentioned substituent in RU486 apparently reacts with a so-called *accessory antagonist subsite,* which normally remains *unfilled* within the *ligand-binding pocket* of the progestin receptor. However, with this additional contact, the receptor remains arrested in an *inactive* or *improper conformation;* thus RU486 acts as an antagonist. Apparently, such a subsite does not exist in progestin receptors of all species.

Thus it can be explained that in birds, RU486 functions as an agonist and not as an antagonist (see pictures in chapter 'DNA-Binding Proteins').

This makes it clear that it is not the structure of the ligand per se that renders it agonistic or antagonistic, but rather the structure of the receptor that determines how a ligand is *interpreted*.

Gestagens, like progesterone, are of great importance in the maintenance of pregnancy. Mutations in the progestin receptor gene lead to habitual spontaneous abortions (loss of embryo/fetus). Substitution therapy of a pregnant woman with gestagens can compensate for the lower affinity for progesterone caused by the mutation and thus may save the embryo. On the other hand, administration of RU486 can terminate a pregnancy. The disease hypercortisolism can also be treated with RU486 since RU486 is, as previously mentioned, also a glucocorticoid receptor antagonist.

Thyroid Hormones

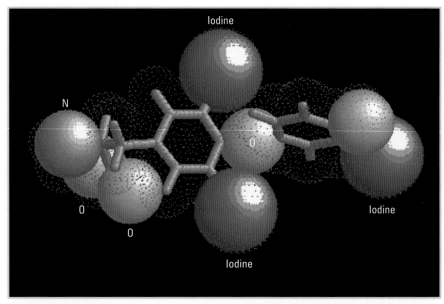

Tri-iodothyronine (T3)

Thyroid hormones are characterized by their possession of 3 (tri-iodothyronine) or 4 (tetra-iodothyronine = thyroxine, T4) iodine atoms, and are derived from the aromatic amino acid tyrosine by conjugation of 2 iodinated tyrosines. Thyroid hormones are not as flat as steroids because they do not have a steroid structure, but are bent at a certain angle in the center of the molecule. In spite of this, their mechanism of action is similar to that of steroids: They react with thyroid hormone receptors (T3R), which, like the steroid receptors, belong to the *superfamily of nuclear receptors*, a subfamily of *zinc finger transcription factors*.

T3 and T4 differ from each other by only one iodine atom, and yet this has enormous biological consequences: More T4 than T3 molecules circulate in the blood because T4 has, by two orders of magnitude, greater affinity for the carrier protein TBG *(thyroxine-binding globulin)* than T3. The opposite is true for its receptor affinity, which is why circulating T4 is regarded as a *prohormone* of T3; the latter is the actual hormone that emerges within the target cell through the action of a *deiodase*.

Dietary iodine deficiency leads to hypothyroidism and if this occurs during pregnancy and remains undetected and untreated, the newborn will be mentally and physically retarded, defects that cannot be compensated for at a later time, but can easily be prevented by timely *substitution therapy* with T3/T4 drugs.

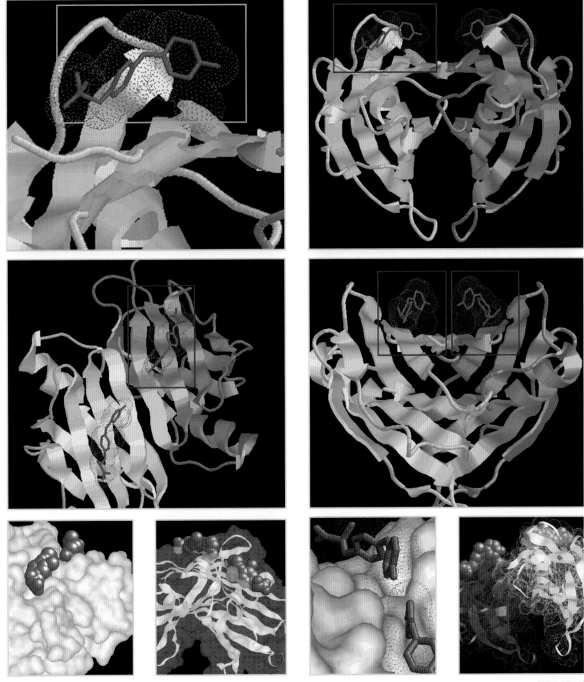

Transthyretin (prealbumin) homodimer **+ T3** in different views (PDB 1THA)

In addition to TBG, other thyroid hormone-binding and -transporting proteins are present in blood, e.g. transthyretin, as depicted here (also called prealbumin). As can be seen, transthyretin is a symmetric homodimer, and each monomer can bind one molecule T3 or T4. A layer of β-sheets forms a 'bed' for the ligand. Several loops, that connect β-sheets, form the homodimerisation contact sites. Whether the sole function of transthyretin is the transport of T3/T4 is as yet not known.

It is interesting to note that a mutation in the transthyretin gene can result in accumulation of pathological transthyretin multimeres and insoluble 'mega' aggregates *(fibrils)*. Such are conspicuous in many slowly progressive neurodegenerative diseases (such as Alzheimer's disease) accompanied by severe mental decline.

Thyroid Hormones

Enzyme Inhibitors

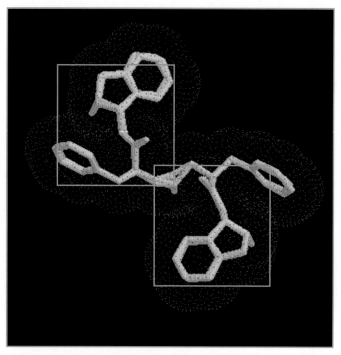

N,N-Bis(2(r)-hydroxy-1(S)-idanyl)-2,6-(R,R)-3diphenyl-methyl-4hydroxy-1,7-heptandiamid in HIV protease

(PDB 4PHV)

Small molecules can consist of rigid halves bent at a precise angle towards each other, halves that may be mirror identical and thus represent, in a certain sense, a single-molecule 'homodimer'. Such a 'double ligand', as pictured here, N,N-Bis(2(R)-hydroxy-1(S)-idanyl)-2,6-(R,R)-3-diphenylmethyl-4-hydroxy-1,7-heptandiamid, can completely fill a binding pocket formed by the homodimerisation of two proteins. Thus, this substance can function as an enzyme inhibitor, preventing the binding of the correct substrate. The ligand depicted here is an inhibitor of the HIV-type 1 protease (PBD 4PHV)(shown to the right in a scaled-down version; for additional pictures, see chapter 'Enzymes'). This particular protease functions similar to a prohormone convertase in that it cuts smaller individual active proteins out of an inactive multiprotein precursor.

This protease is, therefore, essential for the infectious nature and multiplication of the HIV (human immunodeficiency virus), which triggers AIDS (acquired immunodeficiency syndrome).

Inhibition of this enzyme by drugs, like the one shown here, is a part of the current *multidrug strategy* that can render an infected person almost completely virus-free. This drug has been designed from the 3-dimensional knowledge of the enzyme's substrate binding *hole (rational drug design)*, instead of having been discovered by the classical *trial and error* approach.

DNA-Binding Substances
Carcinogens

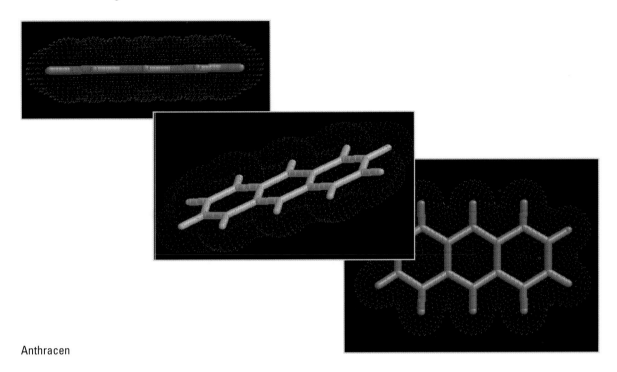

Anthracen

Small, rigid, flat molecules that, like the *anthracen* depicted here and other lipophilic so-called polycyclic hydrocarbons, are built in such a way that they can squeeze themselves between the two strands of DNA.

With the help of highly reactive OH groups, which they additionally receive through cell-internal metabolism (generally after liver passage or within the target cell itself), they can conjugate with the DNA bases and thereby exert a genotoxic, mutagenic and cancerogenic effect. Cancer of the lung and of the urinary bladder epithelium, among others, are direct results of such substances being constituents of tobacco smoke.

Cytostatic Agents

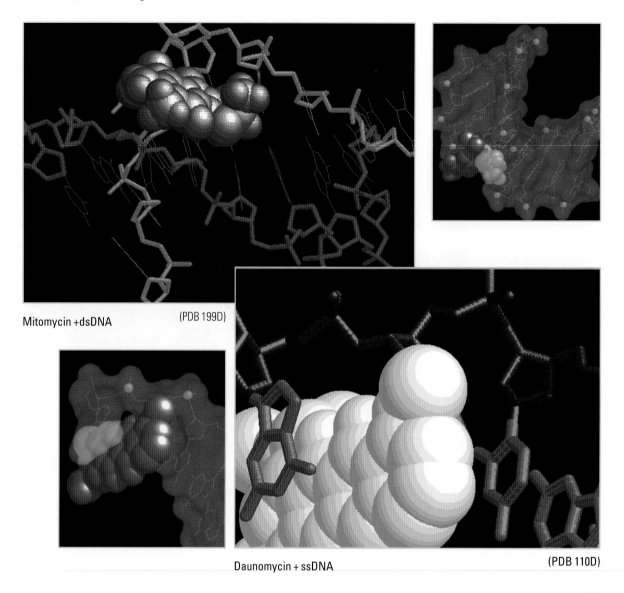

Mitomycin +dsDNA (PDB 199D)

Daunomycin + ssDNA (PDB 110D)

Similarly, other small, rigid and flat molecules, like mitomycin (PBD 199d) depicted here and other cytostatic agents, are built in such a way that they can intrude between the two strands of DNA and, by virtue of a previous metabolic conversion, can react with oppositely lying DNA bases thus covalently intercalating double-stranded DNA.

This would effectively prevent the transient dissociation of the two DNA strands, a prerequisite for *DNA repair, DNA duplication* and *mRNA transcription*. Thus, cells that are ready for division, are arrested in the G2 phase, preventing the terminal steps of mitosis and eventually leading to the *death of the cell*. Many cytostatic agents (*anti-cancer drugs*) are built in such a fashion and react in a similar manner.

Ionophores

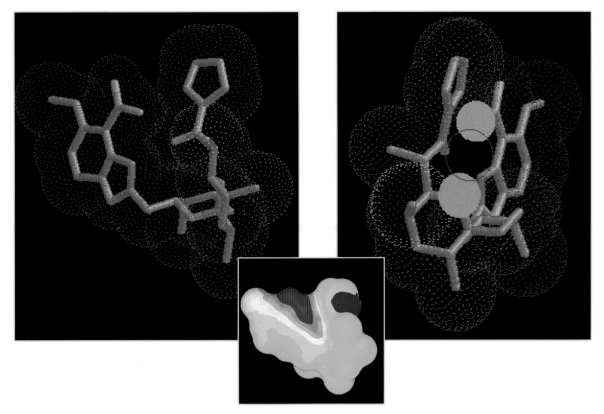

A23187 Ca²⁺-Ionophore, the small picture shows the lipophilic potential of the compound (note central hydrophilic core in red!)

As already mentioned, Ca^{2+} is one of the smallest molecules, a single atom, in fact. Furthermore, a Ca^{2+} ion is positively charged and hydrophilic, and thus membrane-impermeable (only lipophilic substances can pass the cell membrane by simple diffusion). Ions have to be transported through the cell wall either actively, through *ion pumps,* or by a process called *facilitated diffusion,* using *ion transporters.* Furthermore, ions can flow into or out of a cell through *ion channels* along an ion concentration gradient, provided that the channel specific for the ion in question is in an open conformation. All these carriers and channels are complex proteins and thus can be allosterically activated or inhibited by small ligands, i.e., by the endogenous proper ones as well as by exogenous toxins *(opportunistic ligands).* What is true for ions is basically also true for all hydrophilic substances, such as amino acids, sugars, etc. These, too, can be transported into or out of a cell only by membrane proteins of the *solute carrier* superfamily.

If one thinks of a larger molecule, such as a substance designated A23187, which is lipophilic and thus membrane-permeable, but which, under its 'lipophilic cover' possesses an electrically negative and hydrophilic core (center), a Ca^{2+} ion (depicted here as an empty magenta-colored circle) can be held in a chelate-type complex, bound therein by two appropriately positioned oxygen atoms. Thus, Ca^{2+} can be *channeled* through a cell membrane. Lipophilic molecules with hydrophilic cores are called *ionophores*, which display specificity for certain ions, corresponding to the microstructure of their ion-binding cores.

Ionophores are important tools in cell biology research and pharmacology. A23187 is a small synthetic ionophore.

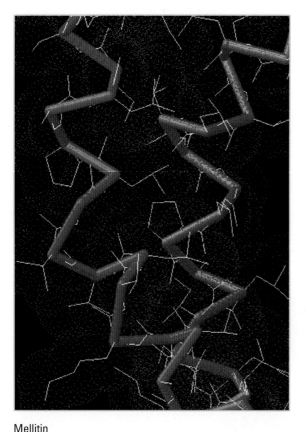
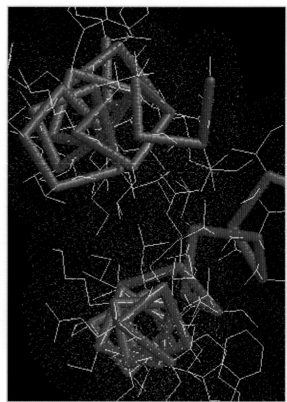

Mellitin (PDB 2MLT)

Also various naturally occurring toxins, such as monesin and valinomycin (a K$^+$-ionophore from *Streptomyces fulvissimus*), are ionophores (just larger ones, relative to A23187) in which a cyclisized chain of lipophilic amino acids forms the coat of the membrane-permeating molecule, and certain hydrophilic amino acids or sugars pointing to the interior of the circle form the ion-coordinating center. Such diffusible toxins that migrate through membranes are called mobile carrier ionophores.

Even larger toxins, such as gramicidin (from *Bacillus brevis*) or mellitin (PDB 2MLT, found in bee venom and depicted above) are lipophilic helical proteins that embed themselves in the membrane, where they can form a mini-ion channel (with more or less specificity for a certain ion). This is toxic for the cell because such a 'primitive' ion channel is unregulatable, and thus persists in a permanently open conformation. Accordingly, such toxins are called channel-forming ionophores. A quite different class of toxins exert their effect by permanently closing a specific ion channel, e.g. bungarotoxin, as will be described later.

The endogenous cell-killing toxins of the human immune system, such as the pore-forming complement components or perforin secreted from cytotoxic T lymphocytes or NK (natural killer) cells, are also based on this channel-forming principle. The above pictures show mellitin as a 2-helix protein, on the left seen from the side, and on the right seen axially from above. It is assumed that mellitin positions itself vertically in the cell membrane, and that a single ion-permeable pore lies between the two helices. Note that the length of the helices is 6 rounds, similar as seen in transmembrane helices of most transmembrane proteins (see chapter Receptors).

Peptidic and Proteohormones

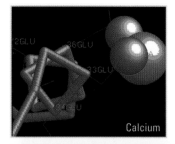

Calcium

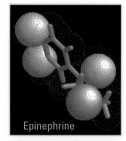

Epinephrine

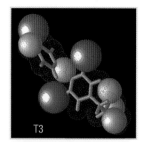

T3

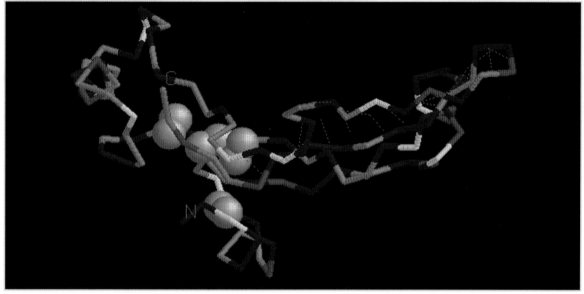

TGFβ (Note the size difference: TGFb is about 300 times larger than a calcium ion!)

(PDB 1TFG)

There are relatively few nonpeptidic small molecule ligands in the signaling system. They include the amine neurotransmitters previously described, such as the catecholamine norepinephrine (NE), epinephrine (E) and dopamine (DA), or serotonin (5-hydroxytryptamine, 5HT), or histamine (H), or the excitatory amino acids glutamate (GLU) and aspartate (ASP), the inhibitory amino acids such as glycine (GLY) and gamma-aminobutyric acid (GABA). All these nonpeptidic neurotransmitters are hydrophilic and therefore act upon cells via membrane receptors that are either of the *metabotropic* 7TMHR superfamily-type or of the *ionotropic*, i.e. ion channel type. In addition, there are the prostaglandins, which are amphiphilic, but still act via 7TMHRs.

Other hydrophilic, nonpeptidic small molecule ligands are the signal transduction molecules inside the cell. To this class of ligands belong cAMP and Ca^{2+}, described above, as well as diacylglycerol, inositol trisphosphate, ceramide and a few others.

To the class of lipophilic nonpeptidic small molecule ligands also belong the steroid and thyroid hormones that, bound to a specific hydrophilic carrier globulin, circulate in blood. Upon arrival at a target cell, they dissociate from the carrier and can diffuse into the target cell, where they interact with a specific receptor that actually is a signal-dependent transcription factor (belonging to the nuclear receptor superfamily). Retinoic acid belongs to this class of ligands. All these mediate predominantly genomic effects, whereas the hydrophilic ligands produce nongenomic as well as genomic effects.

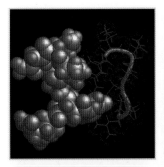
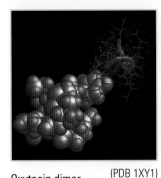

Oxytocin dimer (PDB 1XY1)

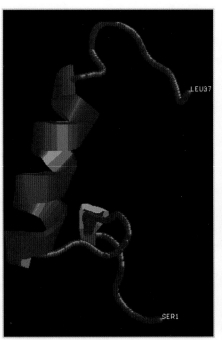

Parathyroid hormone (PTH) 1-37 (PDB 1HPH)

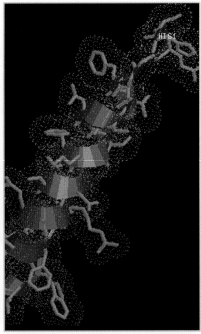

Glucagon 1-29 (PDB 1GCN)

The majority of the *signaling ligands* are, however, peptides that cover a spectrum ranging from the tripeptide TRH (*thyrotropin-releasing hormone*) with a molecular mass of (MG) about 300 to large glycoprotein hormones with a mass of 40,000. Through multimerisation, even larger molecules than that can develop, such as the homotrimer TNFα (*tumor necrosis factor*).

Like the nonpeptidic ligands, peptide ligands also can be grouped together in *families*. Often there is not just one representative of a peptide, but several related analogs. Relationship to a family is based on *similar primary structures*, in which case such members are genuine 'relatives', or may also be assumed from the existence of a *common structural motif* with otherwise little similarity in sequence.

Some representatives of peptide ligands are described below. The structural spectrum ranges from a random coil (unordered) structure, applying to very small peptides such as the tripeptide TRH, the pentapeptide *metenkephaline* (an endogenous opioid) and the nonapeptide *oxytocin*, to highly complex multimers such as TNFα.

(1) Peptides that consist of only one α-helix, such as **parathyroid hormone *(PTH)* or glucagon.**

Excessive secretion of PTH, as a result, for instance, of a mutation in the *calcium-sensing receptor* or caused by a parathyroid tumor, leads to pathological bone loss. PTH deficiency, resulting from surgical removal of the parathyroid glands, reduced activity of PTH due to *PTH gene* mutation, or mutations in the *PTH receptor* gene or postreceptor proteins, leads to life-threatening calcium deficiency, tetanic cramps and numerous other symptoms, such as Albright's hereditary osteodystrophy (AHO).

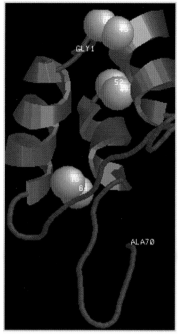
IGF-1 (PDB 1GF1)

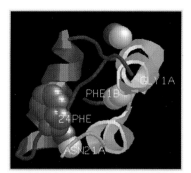
Insulin (PDB 1APH)

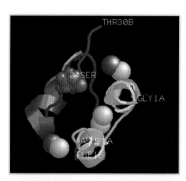
Mutated Insulin (PDB 1HIQ)
'Chicago': PHE24SER

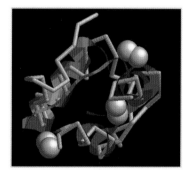

'Chicago'-insulin (grey) and wild-type insulin (magenta)
Note that there is partial overlap but also structural deviations!

(2) Peptides that consist of three relatively short α-helices, such as **insulin** *or* **insulin-like growth factor (IGF-1)**.

Insulin deficiency resulting from an autoantibody and/or autoreactive T lymphocyte-mediated destruction of insulin-producing pancreatic β cells causes diabetes mellitus type I, which, under insufficient and inconsistent treatment, leads to diabetic coma, gangrene of the extremities, blindness, kidney failure and early death. A mutation in the insulin gene that abolishes its hormonal effect can lead to the same phenotypic result.

A similar phenotype can be caused also by blocking autoantibodies against the insulin receptor or by inactivating mutations in the gene for the insulin receptor or for receptor-downstream operating signaling proteins (diabetes mellitus type II, non-insulin-dependent diabetes mellitus, NIDDM). Diabetes mellitus is a very widespread disease!

Lack of IGF-1 resulting from a genetic defect in the HGH receptor gene is the cause of the Laron-type dwarfism, which cannot be treated with growth hormone (HGH) substitution (see chapter on 'Membrane Receptors') but, theoretically, only with IGF-1. Since, as the pictures show, IGF-1 very much resembles insulin, IGF-1 can also react with the insulin receptor and so trigger hypoglycemia, which limits its therapeutic use.

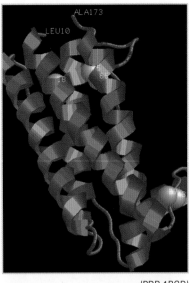
G-CSF (PDB 1BGD)

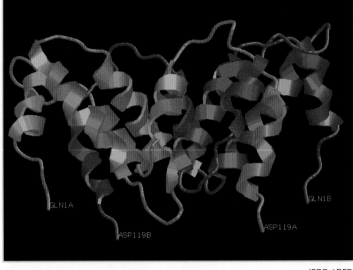
IFNγ (PDB 1RFB)

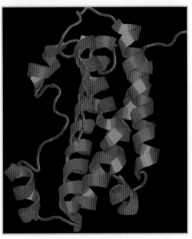
HGH

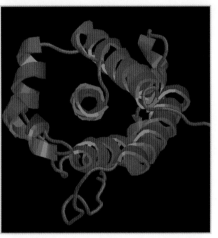
(note here the central α-helix) (PDB 3HHR)

(3) Peptides that are composed of four or five relatively long α-helices parallel to each other (helix bundle proteins), as in the case of the members of the **cytokine/growth hormone family.**

Representatives are the colony stimulating factors (CSFs), interferons (IFNγ, active as homodimers), human growth hormone (HGH or somatotropin) and others.

An inactivating mutation in the HGH gene is responsible for growth retardation, a disease that in earlier times, when left untreated, led to *dwarfism*. Today, such patients can be treated with recombinant HGH, enabling them to grow almost normally. Excessive release of HGH caused by cancerous development of the somatotrophs (i.e. the HGH-secreting pituitary cells) leads to gigantism in childhood or acromegaly later in life, respectively.

Mutations can cause excessive secretion of specific CSFs, which can induce various types of *leukemia*. Mutations can also cause a deficiency or inactivity of specific CSFs, leading to a decrease of leukocytes with the associated consequences, e.g. reduced defense against microbial agents and vulnerability to opportunistic infections. Defect in the macrophage-CSF leads, in mice, to a lack of macrophages and their descendants, such as osteoclasts, which results in diminished or absent *bone remodeling*, i.e. the maintenance of a dynamic steady state by sequential resorption and formation of bone matrix. The disease resulting from a nonfunctional bone remodeling is called *osteopetrosis* or *marble bone disease*. Interestingly, in an equivalent disease in man, a mutation in the carbonic anhydrase type II gene has been discovered.

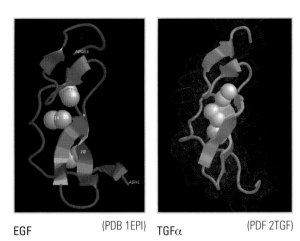

EGF (PDB 1EPI) TGFα (PDF 2TGF) bFGF (PDB 1BFG) IL-1β (PDB 2MIB)

To the right: Superimposition of **bFGF** (magenta) and **IL-1β** (grey). Note the similarity in overall structure! These two factors obviously belong to the same family of proteins.

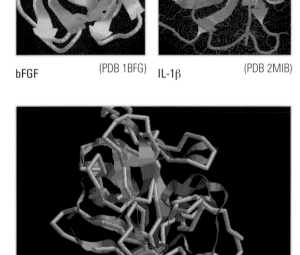

(4) Proteins composed predominantly of β-sheets,

which can be relatively short and parallel or antiparallel-ordered, e.g. in epidermal growth factor (EGF) and its relative, i.e. transforming growth factor alpha (TGFα). They can also display 'propeller' or sandwich (jelly roll) forms, such as in basic fibroblast growth factor (bFGF, 146 AAs) or in interleukin 1β (IL-1β). In tumor necrosis factor alpha (TNFα, active as a trimer), the β-sheets are not arranged in the form of a fan, but are antiparallel to each other (see page 23).

In most of the peptide/protein ligands, the internal disulfide bridges constitute an essential structuring principle (in this book, the γ-sulfur atoms are generally highlighted as CPKs), such as shown in this chapter in IGF-1, insulin, G-CSF, EGF, TGFα, TNFα, NGF, PDGF, TGFβ, hCG).

Mutations can lead to an excessive secretion of growth factors and excessive proliferation (cell division) of cells sensitive to the respective growth factor. In principle, an over-expression of receptors for growth factors functions in the same way. For this reason, growth factors, their receptors as well as the downstream triggered signal transduction proteins are comprehensively designated as *cellular* oncogenes due to their cancer-causing potential when abnormal. There are cellular oncogenes that have, earlier in evolution, been '*shuffled*' to viruses (i.e. incorporated into their genomes), but in incomplete or mutated forms, or have later undergone various mutations that can, by reinfection with a similar (but, of course, never the same) virus, as *viral* oncogenes, be transmitted to healthy normal host cells. As mutated, i.e. abnormal genes, viral oncogenes represent dysregulated or unregulatable or otherwise abnormal oncogenes, whereas normal cellular oncogenes induce only normal, i.e. regulated proliferation of cells.

Over-expression of EGF receptor (EGFR) occurs in many breast carcinomas. Measurement of EGFR is thus important both for diagnostic and prognostic purposes.

FGF has a differentiating and activating effect on various mesenchymal cells. FGF is necessary, among others, for *angiogenesis*, i.e. formation of new blood vessels. Because cancerous tumors require a great deal of oxygen for survival and growth, they induce localized angiogenesis for this purpose. Suppression of angiogenesis may thus be the most effective antitumor therapy! Mutations of growth factors and their receptors can result in their reduced or functionally defective expression and respective biological consequences, e.g. to severe deformation of the bony cranium (acrocephaly), characteristic of Crouzon syndrome, which is due to mutations in several FGF isoreceptors.

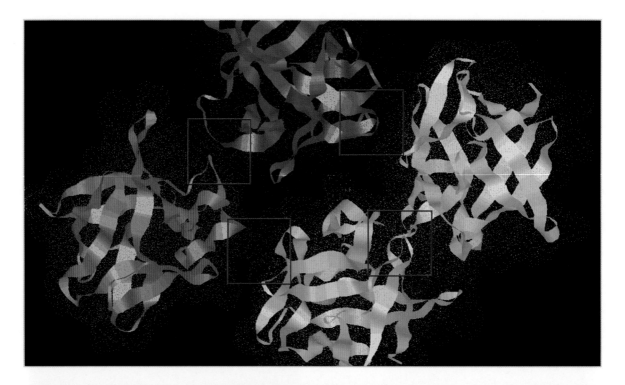

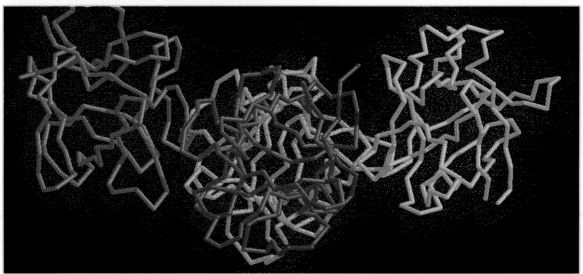

Acidic fibroblast growth factor (aFGF) tetramer (PDB 1AFG)

Acidic fibroblast growth factor (aFGF) occurs as a homotetramer and also acts as such. The top picture, highlighted by frames, shows that every monomer possesses two topographically distinct multimerization sites. Below, the tetramer is turned 90° in the x axis relative to the picture above, and thus shows an equatorial view.

FGF is a rare example of hydrophilic signal substances that, after binding to a membrane receptor, are also endocytosed and even translocate to the nucleus. Their mode of action there, however, remains to be clarified. Tetramerization, at least, appears to be a prerequisite for endocytosis.

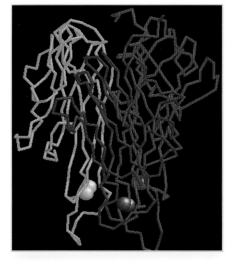
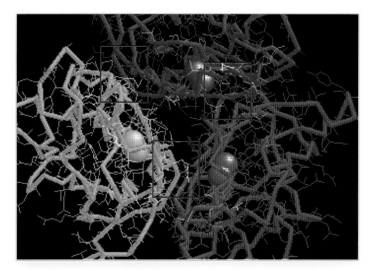

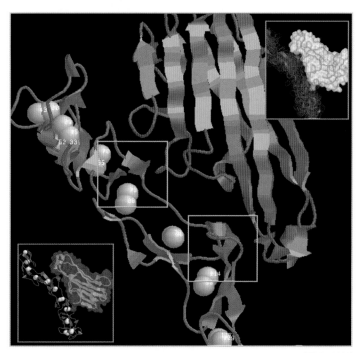

TNFα trimer (PDB 1TNF)

TNFα-monomer complexed with the extracellular domain of the **TNF receptor**: two *contact sites* highlighted by white frames (PDB 1TNR)

Tumor necrosis factor-α (TNFα) is one of the most important signals that induce *programmed cell death (apoptosis)* in many cell types. The top right picture shows, in an axial view of TNFα, that every monomer possesses two multimerization sites. The Fas ligand and Fas receptor, further apoptosis mediators, have analogous structures and are also trimeric.

Apoptosis also requires the presence and functioning and cooperation of many **proteins** (*apoptosis apparatus*), just as cell division (*proliferation apparatus*) as well as every other process in a cell, an organ or an organism does. This means a cell is only able to divide or die, respectively, when it possesses the normal number of normally functioning proteins necessary for these processes at that particular moment. Mutations in the corresponding genes render a cell unable to die or to grow, respectively. The former has been hypothesized to be a cause of *cancer*, in that aberrant *immortality* of cells may eventually lead to their pathological accumulation, just as excessive cell division can.

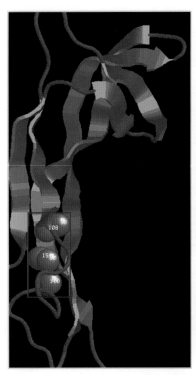 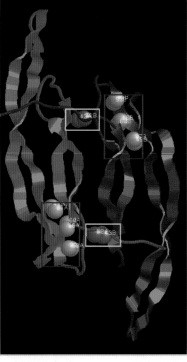 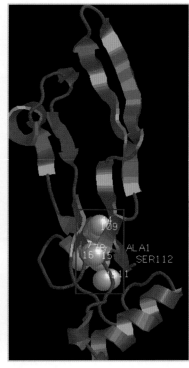

NGF   (PDB 1BET)   **PDGF**-AB-dimer   (PDB 1PDG)   TGFβ   (PDB 1TFG)

(5) The superfamily of the cystin-knot proteins constitutes a group by itself.

The most important representative, TGFβ, has already been described in great detail. TGFβ is the paradigm of a whole family of factors that can induce both cell division and apoptosis. The Mullerian duct inhibiting factor (MIF) belongs to the latter class of factors, that in men suppresses the further development of the Mullerian duct derivates (i.e. fallopian tubes and uterus).

A mutated MIF results in the Mullerian duct persistence syndrome.

Additional members of the cystin-knot superfamily are nerve growth factor (NGF), platelet-derived growth factor (PDGF, a homodimer covalently linked via two intermolecular disulfide bridges) and the subfamily of the glycoprotein hormones comprising, among others, human chorionic gonadotropin (hCG). The latter consist of an α- and a β-subunit that heterodimerize by noncovalent interaction, as mentioned previously and detailed below.

NGF, PDGF, TGFβ and others can exert their biological effect only when multimerized because they have to cross-link two or more neighboring receptors, a requirement typical for growth factors. The glycoprotein hormones, by contrast, bind only to a single receptor but have to be α:β-dimers because their receptor contact domains are formed and completed only by both subunits correctly bound together. Of course, there are also growth factors that, as monomers, possess two receptor contact sites and can simultaneously crosslink two receptors, e.g. HGH, IL-2 and other cytokines (see later).

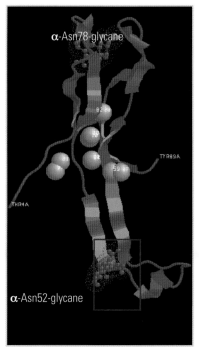 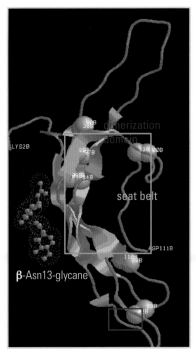 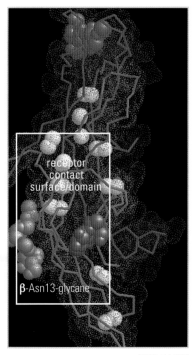

α-Subunit   β-Subunit   α-β-Dimer = HCG   (PDB 1 HRP)

In contrast to most of the other growth factor dimers, hCG is a heterodimer. It consists of an α-subunit with 92 amino acids (AA) and a β-subunit with 142 AAs. Each subunit is coded for by its own gene and each is, by itself, a cystin-knot protein. Both subunits consist exclusively of ß-sheet elements and loops. Similar to PDGF, both subunits associate in opposite longitudinal directions ("head to tail"), but non-covalently. An extremely interesting structure is a loop in the β-subunit (highlighted by a frame) that is held close to the stem of the β-subunit by a disulfide bridge forming a kind of seat belt, under which the lower half of the α-subunit is stuck. The mechanisms of this structurally extremely complex dimerization process is an unresolved puzzle.

In the far right panel, the receptor contact domain is highlighted by a frame: One can see that both subunits contribute to it, and that the Asn52-glycane residue of the α-subunit also participates in receptor interaction.

Within a species, there is one α-subunit, but several different β-subunits; in humans, unique β-subunits are provided for TSH (thyroid-stimulating hormone), FSH (follicle-stimulating hormone), LH (luteinizing hormone) and even seven identical β-subunits for hCG encoded by seven isogenes arranged in tandem on the same chromosome. These β-subunits are so different from each other that they confer hormone specificity on each of the α:β heterodimers, i.e. effectiveness as TSH, FSH, LH or hCG, respectively (hCG and LH are quite similar to each other and there is only one LH/hCG receptor). On the other hand, these different β-subunits are in a certain region (i.e., the heterodimerization domain) also so similar to each other that each of them can associate, head to tail, with the same α-subunit.

Every subunit is glycosylated, i.e. possesses glycane residues at certain AAs (i.e. at asparagine, Asn, and serine, Ser) that are essential for the hormonal effect (shown here are the Asn-glycanes in ball and stick + dots representation). Deglycosylated hCG, as previously mentioned, is a receptor antagonist. HCG, which lacks one or more Ser-glycanes, has a highly reduced biological half-life, i.e. is eliminated from circulation much more rapidly than native hCG.

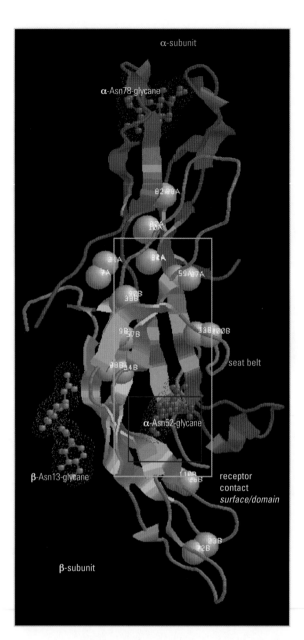

LH and hCG exert their activating effects via membrane receptors on corpus luteum cells in the ovary and on Leydig cells in the testes (stimulating steroid synthesis). FSH exerts its mitogenic and cell differentiation-inducing effects by a similar receptor on the follicle cells of the ovary, and on the Sertoli and seminal tubulus epithelial cells of the testes. The same is true for TSH, which has mitogenic and activating effects on thyroid gland follicle epithelial cells: TSH promotes the uptake of iodine and its incorporation into tyrosine residues of thyroglobulin, from which T3 and T4 can ultimately be produced.

Whether each glycoprotein hormone subunit possesses some biological activity in its monomeric form is yet as not clear, but appears possible.

Inactivating mutations in the various β-subunits lead to delayed/suppressed sexual and thyroid gland development, respectively, and to all of the consequences therefrom [eunuchoid long stature, delayed puberty, azoospermia, anovulation, sterility (infertility), hypothyroidism, cretinism]. If the α-subunit is mutated, all glycoprotein hormones must be inactive, and a syndrome should occur that includes all of the different phenotypes mentioned above.

Recombinant FSH, available today, is employed to induce artificial ovulation in the context of sterility therapy including *in vitro fertilization*. The non-recombinant preparations used earlier were less pure, less specific and prone to result in unwanted multiple pregnancies.

By precise knowledge of all the domains of a hormone and their respective biological functions (receptor recognition, agonist/antagonist property, biological half-life etc.), it is today possible to '*tailor*' proteins by deliberately altering their characteristics for a specific therapeutic purpose ('*designer proteins*').

Peptidic Toxins

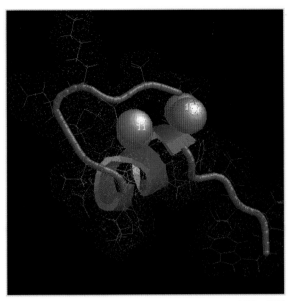

Sarafotoxin (PDB 1SRB)

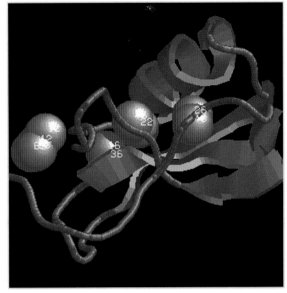

Charybdotoxin (Scorpion Toxin) (PDB 1PTX)

Many toxins, secreted by different species of animals for defense and other purposes, are peptides/proteins. They exert their effect on one or more species, including man, with varying efficiencies (pharmaco/toxicological potencies). Their effectiveness is due to their ability, as *opportunistic ligands*, to interact specifically with a certain protein in the membrane of certain cells of the target species.

No toxin in itself is effective! It becomes so only when a host possesses, by chance (or evolutionary selection), on a certain cell an appropriate target protein that renders this cell susceptible to the toxin, and thus the organism as a whole as well. If this particular protein is evolutionarily highly conserved, it is to be expected that a given toxin is effective in all those species where its target has similar or identical structures.

Interspecies differences resulting from point or larger scale mutation in the course of evolution can, however, render a species 'tolerant', i.e. insensitive to that toxin, because that protein thereby becomes a 'nontarget protein', i.e. unsuitable for that toxin. The same mechanism is also operative when, by mutations in a gene, a protein loses its affinity for an endogenous ligand, e.g. an enzyme for its substrate or a receptor for its hormone, etc.

Chemically, peptide toxins are mostly small in size, i.e., consisting of 20-40 AAs, among which, interestingly, an extraordinarily high number of cysteines occur that, by numerous disulfide bridges, provide a particularly compact and conformationally stable (rigid) structure of the native toxin. Often, this structure also makes the toxins metabolically quite resistant.

Membrane proteins are mostly the target proteins for toxins, e.g. receptors and ion channels on smooth and transverse striated muscle cells or on nerve cells ('neurotoxins'). Why? Because these proteins mediate acute effects such as neurotransmitter release and myo-contractions, which explains the rapid onset of the toxins' effects (i.e. within seconds to minutes). Long-term blockade or inactivation (occasionally also hyperactivation) of such membrane proteins obviously represents the best weapon of a food-hunting animal to render the victim lame or rigid, and thus 'ready' to be killed and eaten. Some species can secrete a veritable *cocktail* of different toxins aimed at different target proteins, thus increasing and potentiating the toxic efficiency, similar to a 'stratified' multidrug strategy as used in human medicine against cancer or infections such as AIDS.

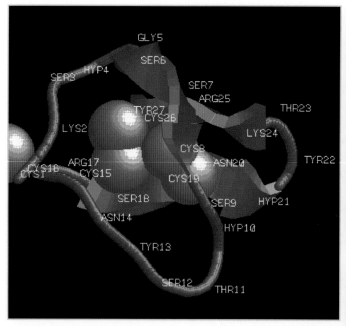

ω-Conotoxin (PDB 1ABT)

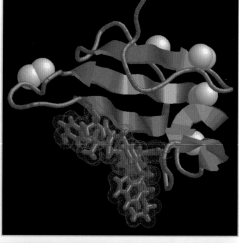

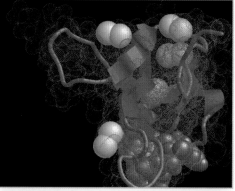

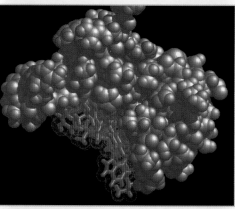

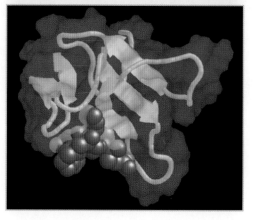

Right row: α-Bungarotoxin complexed with a peptide fragment of the extracellular domain of the nicotinic acetylcholine receptor ion channel (in magenta, wireframe and dots)

Sarafotoxin, from the poisonous snake *Atractaspis enggadensis* (found in Israel) is a high-affinity and long-acting agonist on the *endothelin receptors* in heart atrial myocytes as well as in the central nervous system. *Charybdotoxin* from the scorpion *Leiurus quinquestriatus* blocks *voltage-gated K^+ channels*. Ω-*Conotoxin* GVIA (from the fish-hunting sea snail *Conus geographus*) blocks *voltage-gated N-Type-Ca^{2+} channels*. *Bungarotoxin* (from the cobra *Bungarus multicinctus*) blocks peripheral *acetylcholine-dependent* ('nicotinic') Na^+ *channels* of the motoric end plates, to mention a few. Because of their structural rigidity, many such *neurotoxins* can function as *open channel blockers*, as they recognize a site in the pore of the channel that becomes only accessible in the open state of the channel protein.

Various animals and plants produce – with similar 'intention' – nerve and muscle poisons on a nonpeptidic basis as well, e.g. *tetrodotoxin* (from pufferfish), or *saxitoxin* (from a kind of sea plankton), which blocks *voltage-gated Na^+ channels* in the closed state, or *veratridine* (from a species of lilies), or *batrachotoxin* (from the skin of the Columbian frog *Phyllabates aurotaenia*), which arrests *voltage-gated Na^+ channels* in the open state. *Atropine, tubocurarine, strychnine* and a series of other small molecule-toxins, mostly *alkaloids* (see morphine), should also be mentioned here.

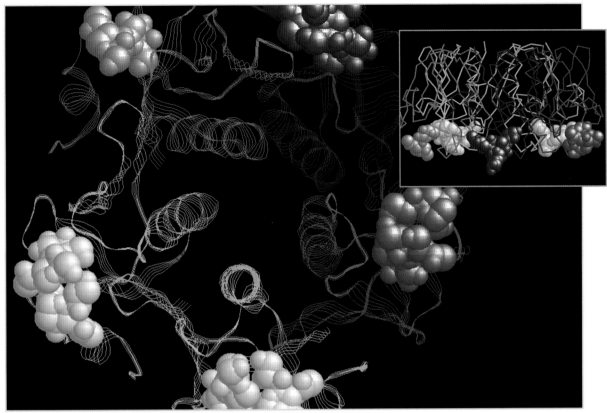

Choleratoxin ß-subunit plus complexed glycanes (inset shows side view)

(PDB 1CHB)

What is common to most of the above-mentioned toxins is that they interact as ligands with extracellularly accessible domains of their membrane target proteins. But there are also toxins that exert their toxic effects intracellularly. For this, they have first to gain entrance into the cell, again by extracellular interaction, by which receptor-mediated endocytosis opens their way. *Choleratoxin*, for example, is an exotoxin of the gram-negative bacteria *Vibrio cholerae*, consisting of a pentamodular 87-kD large β-subunit and a small 27-kD α-subunit. The latter is the actual toxic enzyme that interacts with the α-subunit of the stimulatory GTP-binding protein (Gsα), ADP-ribosylates it and thus abrogates its inherent GTPase function. Thus, bound GTP is no longer hydrolyzed to revert Gsα into its deactivated resting state. When permanently active, Gsα stimulates indefinitely adenylate cyclase (among other target proteins) which, in the intestinal epithelium, results in enormous water secretion and diarrhea and thus life-threatening dehydration.

How does the toxic α-subunit gain entry into the cell? It does so by being complexed to the larger β-subunit which, as a *pentalectin*, can bind promiscuously to glycanes of various compositions. Since the extracellular domains of all membrane proteins are glycosylated (carry one or more glycane moieties), and since membrane proteins are quite adjacent to each other in the membrane, a multilectin of appropriate overall size should be able to *crosslink* several such glycosylated membrane proteins or glycolipids (e.g. ganglioside GM1) via binding to their glycanes and thus trigger receptor-mediated endocytosis, allowing the toxin to gain entrance into the cell.

Pertussis toxin or *diphtheria toxin* enter a cell in a similar way, but arrest other 7TMHR-associated G proteins or block G proteins essential for protein synthesis, respectively.

Many poisons are important molecular tools for pharmacological research or laboratory diagnosis. *Choleratoxin* is used for differentiating between various 7TMHR-associated G-proteins. *Bungarotoxin*, in the radioactive or fluorescent dye-labeled form, is employed as a high-affinity ligand to determine the presence of antibodies to the nicotinic acetylcholine receptor, as present in a severe muscular disease called myasthenia gravis.

Insert

Pictorial Representations of Proteins

A _____ Ligand

Transforming growth factor β (TGFβ)

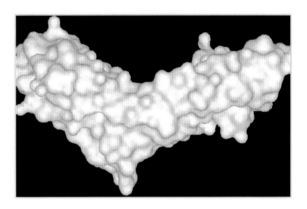

B _____ Ligand-Binding Protein

Human chorionic gonadotropin receptor (HCGR)

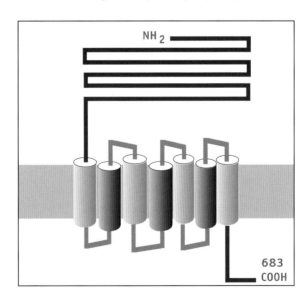

Ligand
Transforming Growth Factor-β

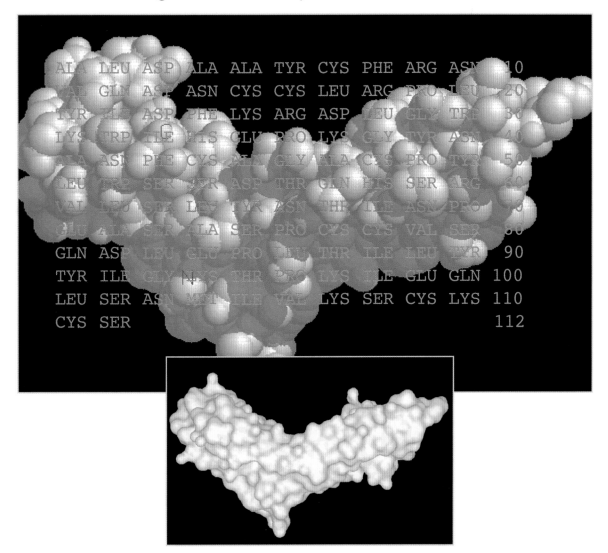

In this series of pictures, different representations of proteins are shown and their characteristic features explained, using TGFβ (PBD 1TFG) as an example. This should provide a basic introduction to the essential nature of proteins in general, since they constitute the main area of interest of this book.

TGFb consists of 112 amino acids (AAs). As shown above, the protein has a long oval shape, resembling a banana or a heart. The picture shown above comes closest to the real appearance of the protein molecule. Here, all atoms of the 112 AAs are depicted as space-filling models (CPKs), each ball corresponding to the individual atoms' van der Waals radius. Of course, only atoms lying on the surface of the protein and on the side facing the viewer can be seen. The inset shows the protein's surface in a smoothed form, as if wrapped in a piece of silk, which corresponds with even greater probability to reality.

How such a conformation of a protein comes into being is hard to guess at from an inspection of its primary structure. The native protein is certainly not a stretched-out string of 112 AAs, but rather is folded, put together, wound up, knotted and compacted. The transformation of a 2-dimensional polypeptide string into a 3-dimensional protein body is based on the inner forces of the polypeptide itself. The requisite information for this folding is laid down in the AA sequence. Important structure-formers are the cysteines.

```
ALA  LEU  ASP  ALA  ALA  TYR  CYS  PHE  ARG  ASN   10
VAL  GLN  ASP  ASN  CYS  CYS  LEU  ARG  PRO  LEU   20
TYR  ILE  ASP  PHE  LYS  ARG  ASP  LEU  GLY  TRP   30
LYS  TRP  ILE  HIS  GLU  PRO  LYS  GLY  TYR  ASN   40
ALA  ASN  PHE  CYS  ALA  GLY  ALA  CYS  PRO  TYR   50
LEU  TRP  SER  SER  ASP  THR  GLN  HIS  SER  ARG   60
VAL  LEU  SER  LEU  TYR  ASN  THR  ILE  ASN  PRO   70
GLU  ALA  SER  ALA  SER  PRO  CYS  CYS  VAL  SER   80
GLN  ASP  LEU  GLU  PRO  LEU  THR  ILE  LEU  TYR   90
TYR  ILE  GLY  LYS  THR  PRO  LYS  ILE  GLU  GLN  100
LEU  SER  ASN  MET  ILE  VAL  LYS  SER  CYS  LYS  110
CYS  SER                                          112
```

If one scans the AA sequence (primary structure) of the TGFβ for cysteines, line by line from top left to bottom right, one is struck by the relatively high number of cysteines present: 9 of 112 AAs. TGFβ is thus referred to as a *cysteine-rich protein*. It can be assumed that 8 of these 9 cysteines form *disulfide bridges* and thus contribute to compacting the protein's structure. It is also noticeable that the cysteines are scattered over the entire sequence. An odd number of cysteines indicates that not all of them are involved in disulfide bindings, leaving at least one cysteine residue free.

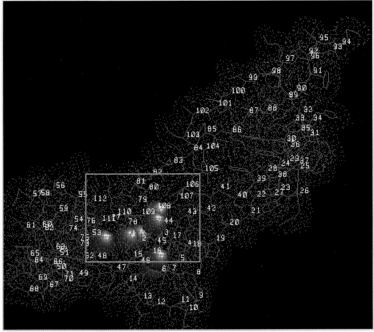

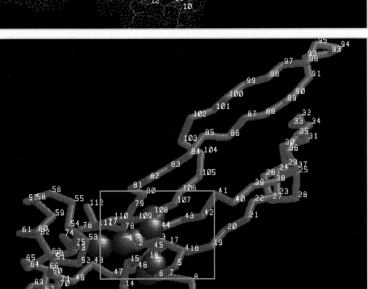

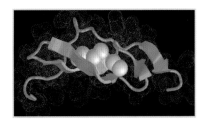

TGFα (PDB 2TGF)

Looking into the architectural plan of a protein shows also that proteins with very similar designations, such as TGFα instead of TGFβ, may be structurally quite dissimilar and thus belong to different protein families. Names have generally a historic basis, often related to the first discovered function, but often bear no relationship to structure.

In the picture above (which, in comparison to the first picture, is turned slightly forward on the X-axis), the total TGFβ is depicted in a dot model representation, i.e. all atoms are shown as transparent 'clouds' of points. The numbers indicate the individual AAs. In the picture below (same viewing angle as above), only the a-carbon backbone of TGFβ is shown, represented as a kind of tube. The AA side chains are omitted. In both pictures, one can also see the g-sulfur atoms of the 8 cysteines involved in disulfide bridges (the 9th free cysteine not being shown) in space-filling (CPK) representation. Each corner point in the backbone represents an a-carbon atom of an individual AA of the polypeptide chain. As can be seen, peptide bonds are always tilted at a specific angle relative to each other. Following the AA numbers, one can trace the folding of the polypeptide string.

Both representations allow the viewer to look, as with an X-ray device, into the interior of the protein, the true surface of which is shown in the first picture, to see the 'skeleton' of the protein and thus understand its architecture.

Thus, the *string* of 112 AAs becomes folded and pleated. As a matter of fact, all 8 of the 9 cysteines are bound in disulfide bridges. Even more surprisingly, all of them that appear scattered in the primary structure are concentrated in a narrowly circumscribed area in the native protein. This structural *motif* is found in a number of other proteins and is called a *cystin knot*.

Ligand 'Transforming Growth Factor-β'

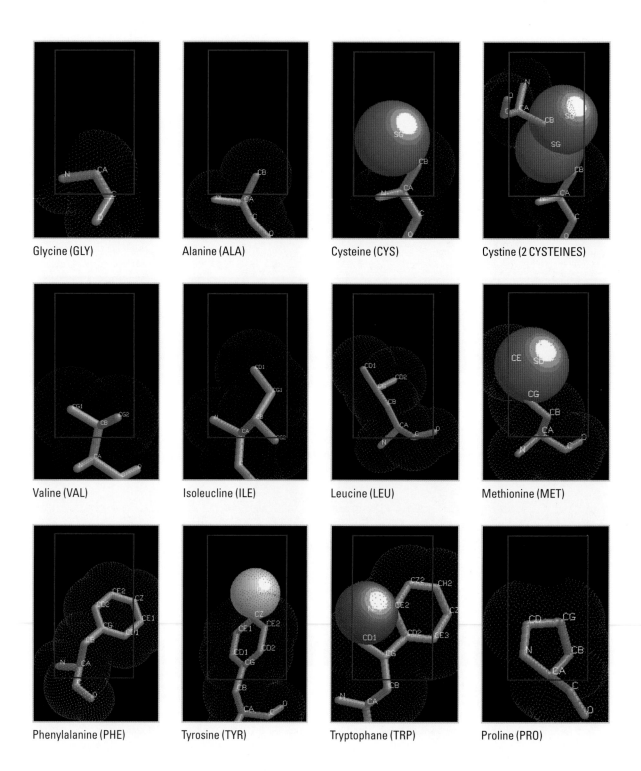

AAs are the building blocks of proteins and there are 20 different AAs. In all pictures here, the Cα (CA) atoms that form the *backbone* of the protein point downwards. Pointing upwards and protruding into the white frame is the so-called *side chain* of a given AA, i.e. the part that confers *size, volume* and *physicochemical character* on the AA.

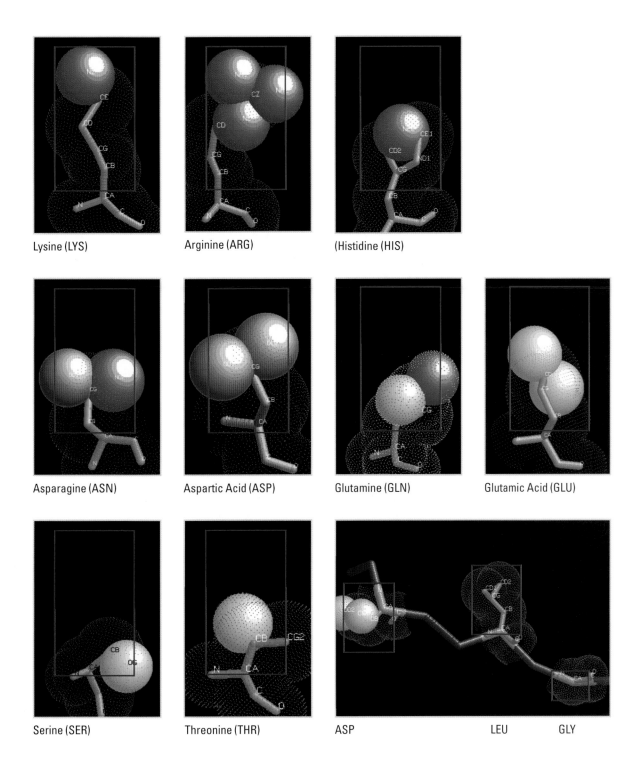

Lysine (LYS) | Arginine (ARG) | (Histidine (HIS)

Asparagine (ASN) | Aspartic Acid (ASP) | Glutamine (GLN) | Glutamic Acid (GLU)

Serine (SER) | Threonine (THR) | ASP LEU GLY

These pictures present the varying sizes, structures and equipment of special atoms of a given AA. Light magenta = Oxygen; grey = nitrogen; magenta = sulfur.

The picture on the bottom right shows that glycine is entirely contained in the protein backbone since it has no side chain, whereas leucine possesses a long hydrophobic side chain (consisting only of carbon atoms), in contrast to aspartic acid that exhibits a relatively long, but hydrophilic, side chain (because of the 2 terminal oxygen atoms) and that is also negatively charged (1 carboxylate anion).

AA on the previous page are mainly hydrophobic ones and on this page hydrophilic ones (compare the table of hydropathicities on page xvii).

Ligand 'Transforming Growth Factor-β'

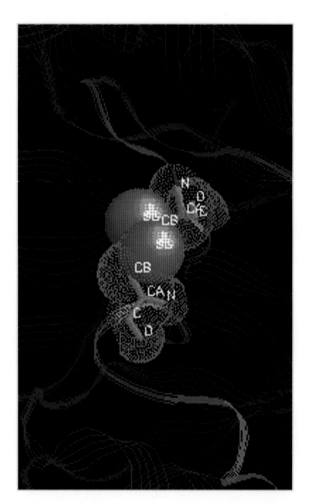
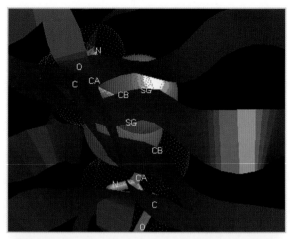
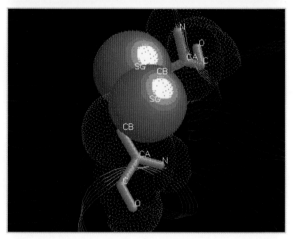

These pictures do not represent TGFβ; instead, they simply show how a large globular protein (of any other kind) can become folded into a compact structure by virtue of a central disulfide bridge. Interestingly, the single disulfide bridge is formed between two cysteines that lay far apart from each other in the primary structure.

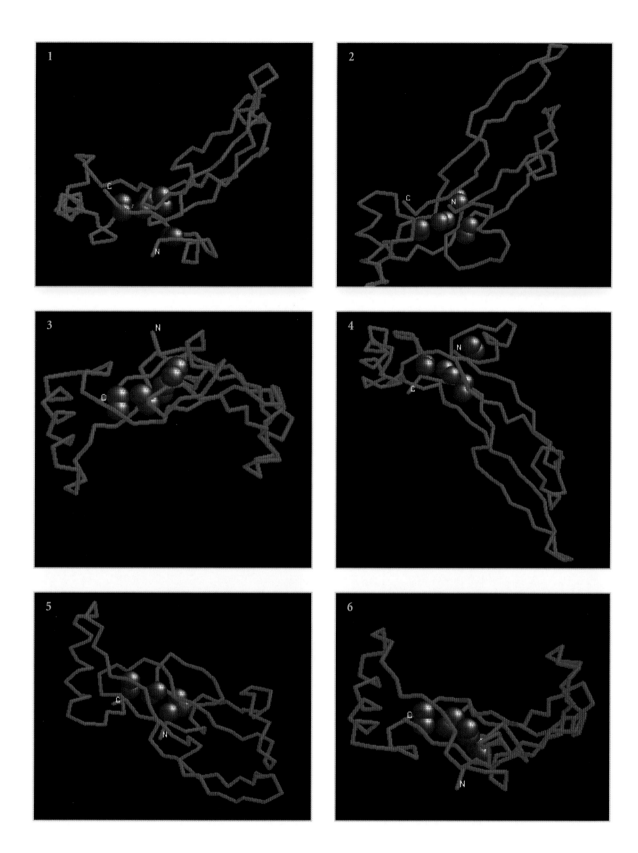

As with a computer tomograph, serial pictures of the backbone of TGFβ (in backbone representation) are taken from different angles. Each successive picture is tilted by approximately 30° in the X-axis relative to the previous picture.

Ligand 'Transforming Growth Factor-β'

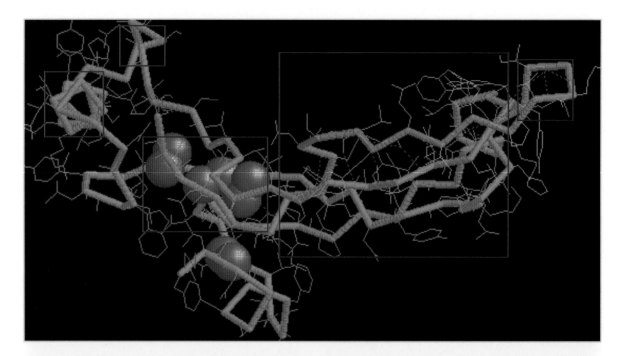

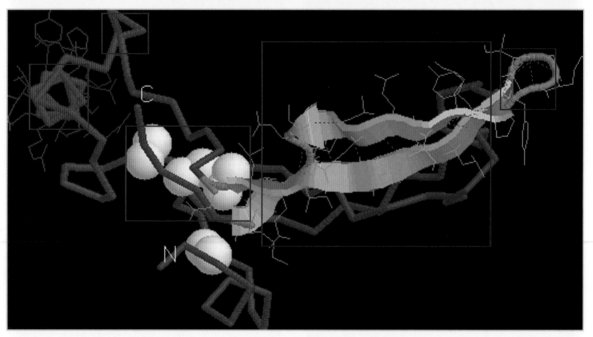

In both of these pictures, taken from the same angle as the first picture on page ii, one can see the backbone plus the individual AA side chains, the latter as wireframes. *Wireframe* lines connect the centers of the constituent carbon, oxygen or nitrogen atoms of the AA side chains. This type of pictorial representation allows a better estimate of the real volume and form of a protein.

The backbone of a protein provides information about the inner architecture and thus gives only an idea of its external shape. The exact overall structure and form is the result of all of the constituent AA residues (or AA side chains) that protrude from the backbone at specific angles. A closer look at the backbone of TGFβ, i.e. by rotating it by one axis (as done in the picture series on page viii), shows that TGFβ consists of the following secondary structure elements: two beta sheets layered above each other (shown to the right in both pictures) and one alpha helix (AA 58-70, shown to the left in the picture below, highlighted in magenta), as well as a few loops (e.g., AAs 92-96). The upper β sheet (AAs 82-109) is shown in the bottom picture, in so-called 'cartoon' representation depicting a smooth silk ribbon strung through the α-C atoms.

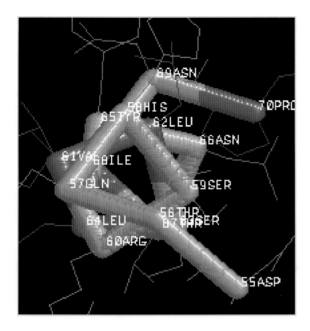
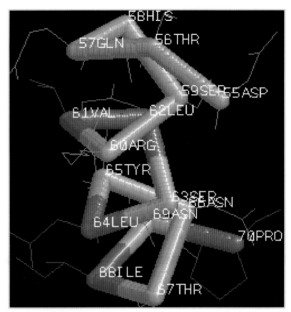
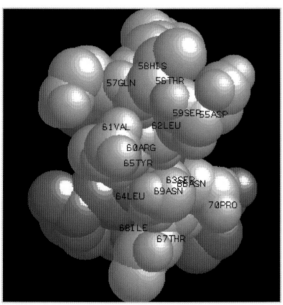
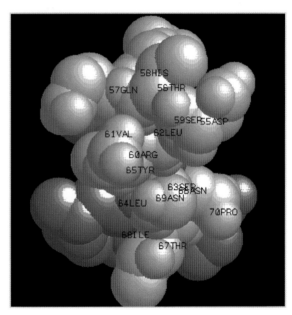

The single alpha helix of TGFβ – 'zoomed in' – an axial view on the left, and a side or longitudinal view on the right; above, in backbone and wireframe representations, below, as space-filling models (CPKs).

It can be clearly seen that the backbone of the chain 55-70 spirals from top to bottom (AA 55 and 70 being outside the helix). Axially seen, the α-C atoms (corners) are in each turn of the coil slightly displaced against each other. Each turn has 3-4 AAs. The complete helix has four turns, which can be easily calculated and recognized from the longitudinal view shown to the right. Further, it can be seen that the individual AA residues of the helix point outwards from the backbone at specific angles relative to the horizontal axis. These angles are determined by geometrical 'rules' of the peptide bonds themselves as well as by mutual interactions of the AA side chains. It should be noted that the central core of the helix is covered, if not filled, with atoms, and does not form an empty channel as might be naively assumed from backbone representations (in the upper panels).

Ligand 'Transforming Growth Factor-β'

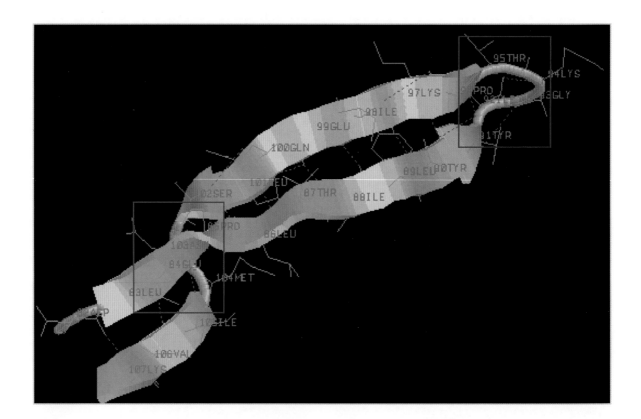

The β sheet 82-107 (cartoon plus wireframe, light magenta) of TGFβ 'zoomed in'. The dotted lines indicate hydrogen bonds between individual AA residues. It can be seen that the β sheet is interrupted by a short piece of loop at AA102, which twists the entire β sheet 82-107 at this site, making the two bands cross each other. AAs 91-95 form a *hairpin loop,* the turning point of the β sheet.

ALA	LEU	ASP	ALA	ALA	TYR	CYS	PHE	ARG	ASN	10
VAL	GLN	ASP	ASN	CYS	CYS	LEU	ARG	PRO	LEU	20
TYR	ILE	ASP	PHE	LYS	ARG	ASP	LEU	GLY	TRP	30
LYS	TRP	ILE	HIS	GLU	PRO	LYS	GLY	TYR	ASN	40
ALA	ASN	PHE	CYS	ALA	GLY	ALA	CYS	PRO	TYR	50
LEU	TRP	SER	SER	ASP	THR	GLN	HIS	SER	ARG	60
VAL	LEU	SER	LEU	TYR	ASN	THR	ILE	ASN	PRO	70
GLU	ALA	SER	ALA	SER	PRO	CYS	CYS	VAL	SER	80
GLN	ASP	LEU	GLU	PRO	LEU	THR	ILE	LEU	TYR	90
TYR	ILE	GLY	LYS	THR	PRO	LYS	ILE	GLU	GLN	100
LEU	SER	ASN	MET	ILE	VAL	LYS	SER	CYS	LYS	110
CYS	SER									112

Scrutinizing the AA sequence (primary structure) of a protein, one can see that AAs with distinct physico-chemical properties are present. Shown here in TGFβ are positively charged AAs in magenta, negatively charged AAs in light magenta, uncharged-hydrophilic AAs in white and hydrophobic AAs in blue.

It becomes apparent that hydrophilic and hydrophobic AAs occur with almost equal frequency and are spread randomly so that from the primary sequence no conclusions regarding the position of an AA in the native protein, its neighboring AAs, a possible clustering of a special kind of AAs, etc. can be made.

Ligand 'Transforming Growth Factor-β'

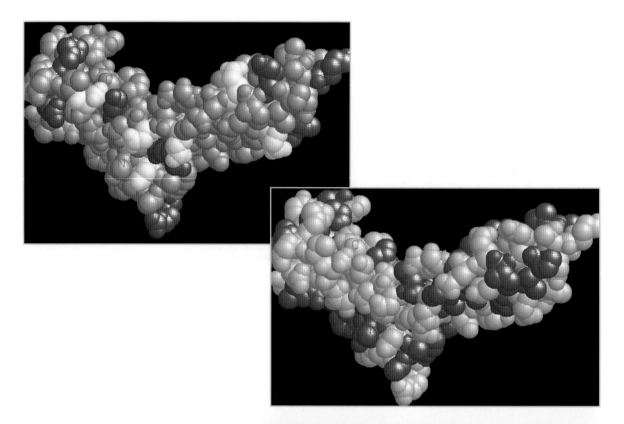

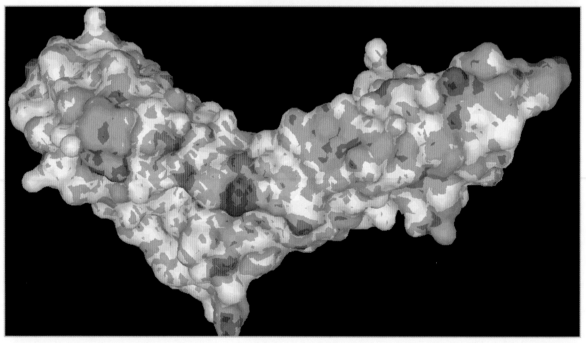

The entire TGFβ in CPK representation. In the top picture, positively charged (Lys, Arg, His: magenta) and negatively charged (Glu, Asp: white) AA residues are distinguished from the rest shown in light magenta. The small picture in the center shows the electrostatic surface in a smooth form, as if the different charges on the molecule's surface shine through a sheet of silk (positive charges blue, negative red).

The picture at the bottom highlights lipophilic AA residues (Ala, Val, Ile, Leu, Cys, Met, Phe, Tyr, Trp) magenta and distinguishes them from the rest of hydrophilic AAs (in light magenta).

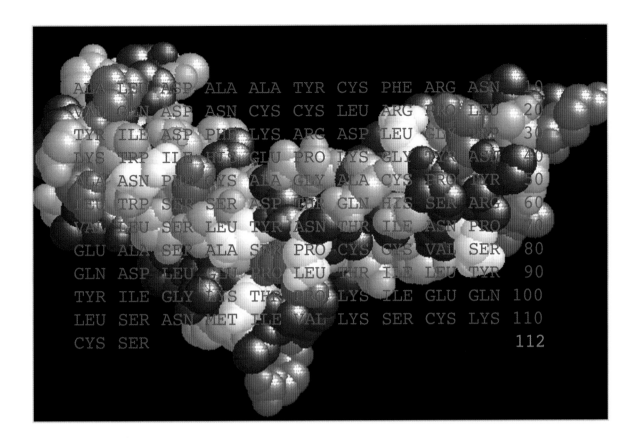

The entire TGFβ in CPK representation. Positively-charged AAs (Lys, Arg, His) magenta, negatively charged (Glu, Asp) white, and hydrophobic AAs (Ala, Val, Ile, Leu, Cys, Met, Phe, Tyr, Trp) dark magenta.

This concluding picture of a series illustrating the different ways of depicting proteins shows that, in addition to the macroform of the complete protein, *microcharacteristics* of *specific surface areas* of the protein are important. That is, a protein's surface can be divided into patches or segments with special hydropathic and/or electrostatic properties. Several such segments can form a larger 'super patch', and the entire surface shows a specific topographic distribution, or *pattern*, of such characteristic patches. Such a super patch may thus constitute a single unequivocal domain, given its specific 3-dimensional form: Rising above the macrosurface or bouncing back from it, as a *ridge*, a *cleft*, a *groove*, a *pocket*, or more generally, as a *'domain'* or a *'site'*.

It is such an ensemble of local characteristics in shape, hydropathicity and charge that permits a protein to interact in specific, spatially precisely defined, unequivocal ways with other proteins, e.g. with specific DNA regions, glycanes, lipids, and/or small ligands. A protein can also interact with the same or another protein: Of course, homo- or heterodi- or even multimerization is also structurally precisely defined (as shown with transthyretin and as will be shown in several later pictures).

Ligand 'Transforming Growth Factor-β'

Ligand-Binding Protein
Human Chorionic Gonadotropin Receptor

```
MKQRFSALQLLKLLLLLQPPLPRALREALCPEPCN
CVPDGALRCPGPTAGLTRLSLAYLPVKVIPSQAFR
GLNEVIKIEISQIDSLERIEANAFDNLLNLSEILI
QNTKNLRYIEPGAFINLPGLKYLSICNTGIRKFPD
VTKVFSSESNFILEICDNLHITTIPGNAFQGMNNE
SVTLKLYGNGFEEVQSHAFNGTTLTSLELKENVHL
EKMHNGAFRGATGPKTLDISSTKLQALPSYGLESI
QRLIATSSYSLKKLPSRETFVNLLEATLTYPSHCC
AFRNLPTKEQNFSHSISENFSKQCESTVRKVSNKT
LYSSMLAESELSGWDYEYGFCLPKTPRCAPEPDAF
NPCEDIMGYDFLRVLIWLINILAIMGNMTVLFVLL
TSRYKLTVPRFLMCNLSFADFCMGLYLLLIASVDS
QTKGQYYNHAIDWQTGSGCSTAGFFTVFASELSVY
TLTVITLERWHTITYAIHLDQKLRLRHRILIMLGG
WLFSSLIAMLPLVGVSNYMKVSICFPMDVETTLSQ
VYILTILILNVVRFFIICACYIKIYFAVRNPELMA
TNKDTKIAKKMAILIFTDFTCMAPISFFAISAAFK
VPLITVTNSKVLLVLFYPINSCANPFLYAIFTKTF
QRDFFLLLSKFGCCKRRAELYRRKDFSAYTSNCKN
GFTGSNKPSQSTLKLSTLHCQGTALLDKTRYTEC
```

The functionality of a protein is based on its intact 3-dimensional structure. Its elucidation becomes difficult, even impossible, when the protein possesses large hydrophobic domains and thus resists crystallization. Attempts have long been made to extract clues from the primary structure regarding the possible ('putative') native structure of a protein. This kind of approach is presented below using a membrane receptor as an example; analogous procedures can be applied to thousands of other proteins. Often the primary structure is deduced from the base sequence of the cDNA coding for the respective protein, since AA sequence analysis is technically more difficult than DNA base sequencing.

The picture above shows the AA sequence of the hCG receptor (hCGR) (AA in single letter code). Clearly, no order and, consequently, no structure is recognizable at first from this type of information.

```
MKQRFSALQLLKLLLLLQPPLPRALREALCPEPCN
CVPDGALRCPGPTAGLTRLSLAYLPVKVIPSQAFR
GLNEVIKIEISQIDSLERIEANAFDNLLNLSEILI
QNTKNLRYIEPGAFINLPGLKYLSICNTGIRKFPD
VTKVFSSESNFILEICDNLHITTIPGNAFQGMNNE
SVTLKLYGNGFEEVQSHAFNGTTLTSLELKENVHL
EKMHNGAFRGATGPKTLDISSTKLQALPSYGLESI
QRLIATSSYSLKKLPSRETFVNLLEATLTYPSHCC
AFRNLPTKEQNFSHSISENFSKQCESTVRKVSNKT
LYSSMLAESELSGWDYEYGFCLPKTPRCAPEPDAF
NPCEDIMGYDFLRVLIWLINILAIMGNMTVLFVLL
TSRYKLTVPRFLMCNLSFADFCMGLYLLLIASVDS
QTKGQYYNHAIDWQTGSGCSTAGFFTVFASELSVY
TLTVITLERWHTITYAIHLDQKLRLRHRILIMLGG
WLFSSLIAMLPLVGVSNYMKVSICFPMDVETTLSQ
VYILTILILNVVRFFIICACYIKIYFAVRNPELMA
TNKDTKIAKKMAILIFTDFTCMAPISFFAISAAFK
VPLITVTNSKVLLVLFYPINSCANPFLYAIFTKTF
QRDFFLLLSKFGCCKRRAELYRRKDFSAYTSNCKN
GFTGSNKPSQSTLKLSTLHCQGTALLDKTRYTEC
```

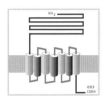

AA sequence of the HCG receptor

From the hydropathicity analysis according to Kyte and Doolittle, one can make a first approximative spatial division of the primary structure of the HCG receptor protein:

The beginning of the first transmembrane helix divides the receptor in roughly two halves: An N-terminal (magenta) domain and a C-terminal (black) domain. The former represents the extracellular domain, the latter is subdivided into the transmembrane domain (TMD, black) and the intracellular domain (ICD, magenta).

```
MKQRFSALQLLKLLLLLQPPLPRALREALCPEPCN
CVPDGALRCPGPTAGLTRLSLAYLPVKVIPSQAFR
GLNEVIKIEISQIDSLERIEANAFDNLLNLSEILI
QNTKNLRYIEPGAFINLPGLKYLSICNTGIRKFPD
VTKVFSSESNFILEICDNLHITTIPGNAFQGMNNE
SVTLKLYGNGFEEVQSHAFNGTTLTSLELKENVHL
EKMHNGAFRGATGPKTLDISSTKLQALPSYGLESI
QRLIATSSYSLKKLPSRETFVNLLEATLTYPSHCC
AFRNLPTKEQNFSHSISENFSKQCESTVRKVSNKT
LYSSMLAESELSGWDYEYGFCLPKTPRCAPEPDAF
NPCEDIMGYDFLR
```

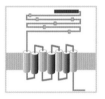

The N-terminal half of the AA sequence of the HCG receptor, representing the extracellular domain of the receptor.
In magenta is the signal peptide, which does not appear on the mature receptor protein since it is cleaved away by a specific peptidase.
C-terminal to the signal peptide comes the actual extracellular domain (ECD). Here again, specific AA motifs, particularly N-glycosylation sites, i.e. asparagine (= N), which, at their free amino groups, are N-glycosylated, are emphasized by magenta. Yet, not every asparagine is glycosylated, i.e. only those localized in a specific context of surrounding AAs: C-terminally to an asparagine, there must be the sequences LS or ES or PS or GT or KT. A so-called N-glycosylation consensus motif or signal thus has the sequence NXS or NXT, whereby X can be any AA.

		Charge	Hydropathicity Index	Gene Codons	Side Chain Atoms
I	Ile		+4.5	ATT, ATC, ATA	4C
V	Val		+4.2	GTT, GTC, GTA, GTG	3C
L	Leu		+3.8	CTT, CTC, CTA, CTG, TTA, TTG	4C
F	Phe		+2.8	TTT, TTC	7C
C	Cys		+2.5	TGT, TGC	C, S
M	Met		+1.9	ATG	3C, S
A	Ala		+1.8	GCT, GCC, GCA, GCG	C
G	Gly		+0.4	GGT, GGC, GGA, GGG	-
T	Thr		-0.7	ACT, ACC, ACA, ACG	2C, O
W	Trp		-0.9	TGG	9C, N
S	Ser		-0.8	TCT, TCC, TCA, TCG, AGC, AGT	C, O
Y	Tyr		-1.3	TAT, TAC	7C, O
P	Pro		-1.6	CCT, CCC, CCA, CCG	3C
H	His	+	-3.2	CAT, CAC	4C, 2N
E	Glu	−	-3.5	GAA, GAG	3C, 2O
Q	Gln		-3.5	CAA, CAG	3C, N O
D	Asp	−	-3.5	GAT, GAC	2C, 2O
N	Asn		-3.5	AAT, AAC	2C, N O
K	Lys	+	-3.9	AAA, AAG	4C, N
R	Arg	+	-4.5	CGT, CGC, CGA, CGG, AGA, AGG	4C, 3N
STOP				TAA, TAG, TGA	

AAs are the building blocks of proteins. Proteins consist of 100 to 1.000 and even more AAs. The notation of the AA sequence employed today uses the the single-letter code (left column, next to it is the well-known three-letter code).

As has been explained in detail earlier, AAs differ in the *atomic composition*, *structure*, *size*, *volume*, and *charge* of their side chains (right column), which gives every AA its specific physicochemical character. Charge is indicated by a '+' or a '−'. The general physicochemical character, which can be more *hydrophilic* (polar) or more *lipophilic* (apolar), can be numerically given by the hydropathicity index. The numbers shown here were taken from Kyte and Doolittle. Isoleucine (Ile, I) is the most hydrophobic of all AAs, arginine (Arg, R) the most hydrophilic.

The global physicochemical character of an entire protein is, simply put, the sum of the characters of the individual constituent AAs or, more precisely, those AAs which, in the tertiary structure of the protein, are oriented such that their side chains become *surface exposed*.

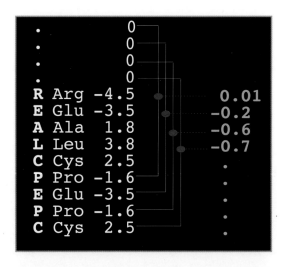

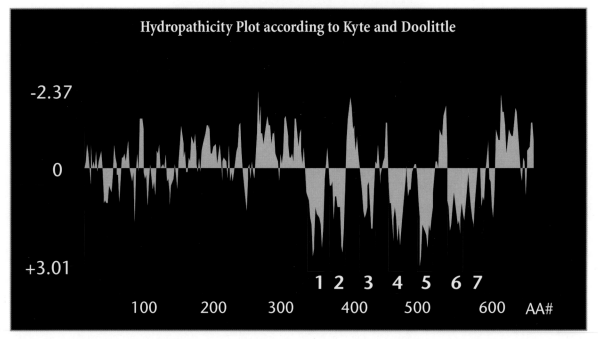

In this example (top picture), the first 9 AAs of the hCG receptor are written below each other. For each AA, its current mean hydropathicity value is calculated as the mean of the individual hydropathicity value of an AA (per Kyte and Doolittle's scale) and the four neighboring N- (lying *above*) and four C-terminal (lying *below*) AAs. Thus, the mean hydropathicity value for each AA is calculated in the context of a 'window' of 9 surrounding AAs. The individual hydropathicities are given in white and the current mean values in magenta.

In a Kyte and Doolittle plot (picture below), the average hydropathicity values (Y-axis) of the individual AAs (676 AAs, X-axis) of the HCG receptor, from N-terminal to C-terminal are plotted. The peaks pointing upwards (-) indicate hydrophilic areas, while those pointing downwards (+) are hydrophobic.

As can be seen, there are seven hydrophobic areas in the C-terminal half of the protein (emphasized by white frames, numbered 1–7). These seven hydrophobic areas have roughly equal widths: about 23 AAs long. They probably represent seven equally long, hydrophobic alpha helices that stretch through the plasma membrane as proven in the case of rhodopsin (see chapter on 'Membrane Receptors').

1. TMH	VLIWLINILAIMGNMTVLFVLLT
1. icl	SRYKLTVPR
2. TMH	FLMCNLSFADFCMGLYLLLIASV
1. ecl	DSQTKGQYYNHAIDWQTGSGC
3. TMH	STAGFFTVFASELSVYTLTVITL
2. icl	ERWHTITYAIHLDQKLRLRH
4. TMH	RILIMLGGWLFSSLIAMLPLVGV
2. ecl	SNYMKVSICFPMDVETTLSQ
5. TMH	VYILTILILNVVRFFIICACYIK
3. icl	IYFAVRNPELMATNKDTKIAKK
6. TMH	MAILIFTDFTCMAPISFFAISAA
3. ecl	FKVPLITVTNSK
7. TMH	VLLVLFYPINSCANPFLYAIFTK

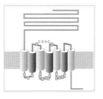

A portion of the hCG receptor's AA sequence, i.e. the transmembrane domain.

This can be further subdivided on the basis of the hydropathicity analysis in an alternating sequence of hydrophobic (magenta frame) and hydrophilic sections.

The seven transmembrane helices (TMH) comprise the hydrophobic stretches, the hydrophilic portions comprise the loops that hold the helices both inside (intracellular loops, icl) and outside the cell (extracellular loops, ecl). Further, it can be noted that positively-charged AAs occur surprisingly more often in the icls (emphasized by magenta) than elsewhere. Furthermore, there is a cysteine (C) both in ecl-1 and ecl-2 (magenta) that, by means of disulfide bridging, pulls TMH-4 and TMH-5 closer to TMH-2 and TMH-3 ('helix packing').

```
VLIWLINILAIMGNMTVLFVLLT  1.TMH
FLMCNLSFADFCMGLYLLLIASV  2.TMH
STAGFFTVFASELSVYTLTVITL  3.TMH
RILIMLGGWLFSSLIAMLPLVGV  4.TMH
VYILTILILNVVRFFIICACYIK  5.TMH
MAILIFTDFTCMAPISFFAISAA  6.TMH
VLLVLFYPINSCANPFLYAIFTK  7.TMH
```

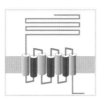

A portion of the hCG receptor's AA sequence, i.e. the seven transmembrane helices. Highlighted in magenta are lipophilic residues that are positionally-conserved in more than one TMH. Conserved AAs of hydrophilic nature are shown in blue.

Such positional conservation of certain AAs suggests that the individual helices had their origin in one common ancestral gene for one transmembrane helix, which, by gene duplication/multiplication and mutation of the replicate gene, developed further. At a later time in evolution, these altered replicates could have been fused into a gene block of seven segments and combined with other genes (i.e. the hydrophilic loops).

```
TFQRDFFLLLSKFGCCKRRAELYRRKDF
SAYTSNCKNGFTGSNKPSQSTLKLSTLH
CQGTALLDKTRYTEC
```

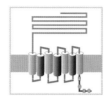

```
TFQRDFFLLLSKFGCCKRRAELYRRKDF
SAYTSNCKNGFTGSNKPSQSTLKLSTLH
CQGTALLDKTRYTEC
```

Portions of the AA sequence of the hCG receptor, i.e. the intracellular domain (ICD).

Remarkably, positively charged AAs (highlighted in magenta) are overrepresented in the ICD (left panel) similar to the **icl**s. The ensemble of the three **icl**s and the ICD, particularly these positively charged AAs within them, are responsible for proper *coupling* of the receptor with signal transduction proteins, i.e., with the α-subunit of the GTP-binding stimulatory protein (Gsα, see pictures later).

Furthermore, as can be seen from the right panel, the ICD also contains quite a few threonines (T) and serines (S) (highlighted in magenta) that, when present in a specific context (similar to the N-glycosylation sites in the ECD), are sites for protein phosphorylation. Protein kinases inside the cell can transiently associate with the ICD and thus phosphorylate these sites. Tyrosine (Y), too, can be phosphorylated, but by a different kinase. Such covalent modifications modulate the functionality of the receptor, primarily leading to a reduced affinity for the agonist (termed '*retrogade conformational change*' or '*desensitization*').

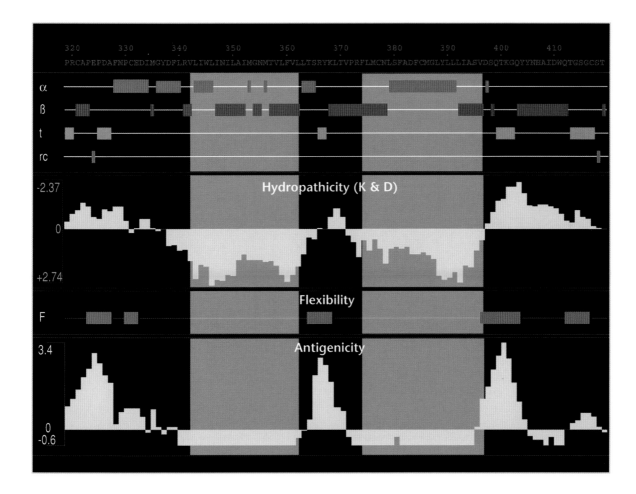

This picture is an enlargement of a portion from the Kyte and Doolittle plot of the entire hCGR, i.e. the first and second *putative* transmembrane helices. Between the two, there is a short hydrophilic stretch (pointing upwards) that represents the first intracellular loop. Downward-oriented are the AAs 342 (V = Val) to 361 (V), which constitute the first transmembrane helix.

In the lower panel, these AAs are presented not in a linear fashion, but rather in a *helical wheel projection* that gives an axial view of the helix. To the right, the same AAs are seen from the side. AAs directed to the front, i.e. towards the viewer, are magenta, those on the back are white. The protein backbone is represented as a grey thread.

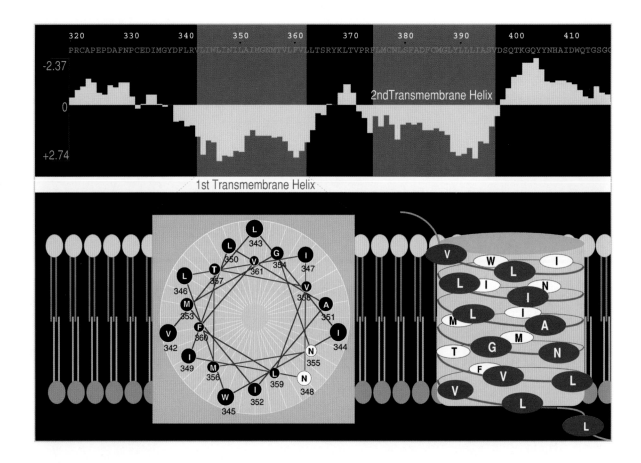

There are further ways to obtain information about the possible structure of a protein from analysis of its AA sequence, as again using the hCG receptor as an example. Algorithms described by Chou and FASSMANN and Garnier and ROBSON, permit estimation of the *probability* that a given AA in a protein is a part of an α-helix or a β-sheet or of a *turn* or a *random coil*, as shown in the first four plots of this picture (α = α-helix, β = β-sheet, t = turn, rc = random coil). It is somewhat disappointing that in this particular example the prediction for the 1st transmembrane helices to form an α-helix is apparently not very high.

The 5th plot once again shows the Kyte and Doolittle hydropathicity diagram, as previously shown. Below are two further diagrams that predict the *flexibility* and *antigenicity* of a protein region, respectively. Hydrophilic portions have high probability to be antigenic, in contrast to lipophilic ones such as the transmembrane helices, whose probability of being flexible and antigenic is low. This appears, however, quite plausible considering they are α-helices that, by virtue of their lipophilic character, are embedded in the plasma membrane and, through *helix packing*, interact with some of the other six helices and thus must be *constrained* in their mobility.

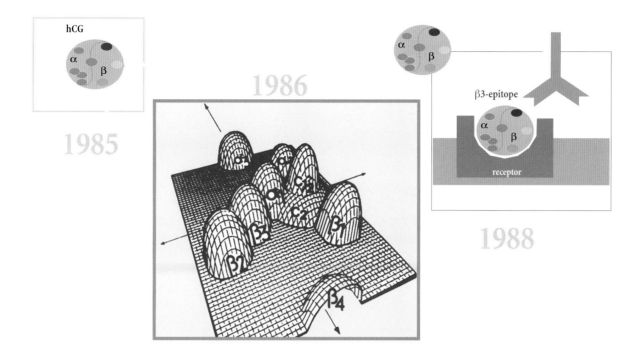

Epitope Mapping in the Author's Laboratory

Elucidation of the crystal structure of a protein is an achievement mainly of the past decade. It has always been clear that unravelling the structure of a protein would be the key to understanding its function. Hence, a variety of experimental methods were developed for this purpose, tacitly accepting that they could, at best, only make indirect statements about the protein's structure. The crystal structure is the *gold standard*, yet it is not always possible to meet it.

A method chosen by the author of this book was ***epitope mapping*** of a hormone such as **hCG**. By using a panel of different *monoclonal antibodies* developed by my colleague Peter BERGER, in 1985, we were able to distinguish – in cooperation with Peter – 14 epitopes on the surface of free hCG (presented schematically in simplified form in the left panel). Five of the epitopes (grey) were positioned on the α-subunit, five epitopes on the β-subunit (magenta), and four were formed by the combination of the two subunits, and consequently did not exist on either of the isolated subunits. Not all of the epitopes are visible on this picture, some must be imagined on the back of hCG. Consequently, the author chose the Mercator's projection, which represents a spread-out form of the cylinder projection of a ball (middle panel), as used in nautical charts (middle panel).

Knowing the number and topographic distribution of the epitopes on the free hCG, I asked my students Elisabeth Nelböck, Heiko Krude and Clemens Lottersberger to address the question of how many and which of these 14 epitopes would still be accessible after hCG was bound by the hCG receptor. To our surprise, we found that only one of the 14, the β3-epitope, was accessible. We called this approach the '**epitope accessibility paradigm**'. From this finding we concluded that the receptor forms a *deep pit* in which hCG submerges. Thus, the greatest part of the ligand's surface becomes masked, and its epitopes inaccessible. Implicit here was the unproven assumption that the pit lays in the level of the plasma membrane (upper panel).

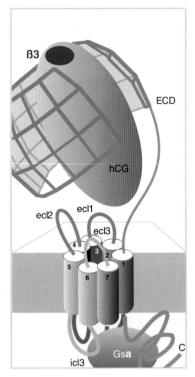

```
                    RELSGSRCPEP
 1    CDCAPDG--ALRCPGPRAGLARLSLTYLP-
 2    VKVIPSQ--AFRGLNEVV-KIEISQSDS-
 3    LERIEAN--AFDNLLNLS-ELLIQ-NTKN-
 4    LLYIEPG--AFTNLPRLK-YLSIC-NTG-
 5    IRTLPDV--TKISSSEFNFLEICDNLH--
 6    ITTIPGN--AFQGMNNESVTLKLYGNG--
 7    FEEVQSH--AFNGTTLIS--LELKENIY--
 8    LEKMHSG--AFQGATGPS-ILDISSTK--
 9    LQALPSH------GLESIQTLIALSSYS--
10    LKTLPSK-EKFTSLLVATL----------
11    --TYPSHCCAFRNLPKKEQNFSFS------
12    ---------IFENFSKQCESTVRKAD-—
13    NETLYSA--IFEENELSGWDYDYGFCSPKT
14    LQCAPEP-DAFNPCEDIMGYAFLR-----

      LzxLPSx--AFzxLxxLx-xLDLSxNx
```

Leu-rich repetitive motif consensus sequence

1989   1992

Further clarification could be achieved only after the hCG **receptor was cloned** in 1989 by Rolf Sprengel in the laboratory of Peter Seeburg, Heidelberg. These authors showed that the hCG receptor belongs to the seven transmembrane helix receptor (7TMHR) superfamily, but differs from them in that it possesses a long = large ECD. This probably represents the putative ligand-binding domain, considering that the other 7TMHRs that bind small ligands such as catecholamines have only a very short — small ECD.

Our findings on the *accessibility of epitopes*, i.e. that the greatest part of the hCG surface disappears upon receptor binding, clearly supported this hypothesis, although the structure of the ECD was totally unknown at this time. Based on the finding that the ECD consists of 14 (a number merely coincident with the 14 epitopes on hCG!) *repetitive leucine-rich modules* (see the sequences above), a characteristic that can also be seen in a series of soluble proteins, in 1992 we drew an intuitive picture of the ECD as a ladle-shaped structure, the inside of which could provide the appropriate space for harboring the ligand hCG.

Following the cloning and deduction of the primary sequence of the hCG receptor, it became clear that the 1985 assumption that the receptor would form a ligand-binding '*pit*' at the level of the plasma lipid bilayer membrane was wrong. Instead, the insight supported the concept that the space at the membrane level formed by the seven transmembrane helices is adequate to take up a small ligand, such as epinephrine or retinal, but not a voluminous ligand, such as hCG.

Thus, we became convinced that it is the recently in evolutionary terms, acquired long ECD that must be considered as *the* ligand-binding domain.

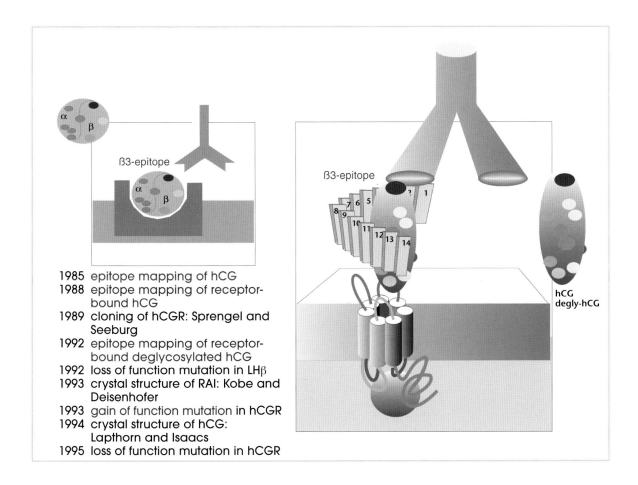

1985 epitope mapping of hCG
1988 epitope mapping of receptor-bound hCG
1989 cloning of hCGR: Sprengel and Seeburg
1992 epitope mapping of receptor-bound deglycosylated hCG
1992 loss of function mutation in LHβ
1993 crystal structure of RAI: Kobe and Deisenhofer
1993 gain of function mutation in hCGR
1994 crystal structure of hCG: Lapthorn and Isaacs
1995 loss of function mutation in hCGR

We carried our investigations further by asking whether **deglycosylated hCG** (degly-hCG) (kindly provided by Wolfgang MERZ, Heidelberg) also possessed the 14 epitopes, just as native hCG. Heiko Krude in my laboratory was able to show that this was, indeed, the case, and we came to the conclusion that the glycane moieties of a glycoprotein hormone, such as hCG, are not integral parts of immunologic epitopes. Degly-hCG is known to be a competitive **antagonist** to hCG on the hCG receptor. It was therefore logical to extend our 'epitope accessibility paradigm'. We showed that, in the receptor-bound state, the β3-epitope as well as the other 13 epitopes became inaccessible (= masked). This was quite a novel finding, as we 'visualized' a **sterically distinct interaction of an antagonist vs. an agonist**. It appears that degly-hCG, relative to hCG, may slip a bit further down (closer to the membrane), where the β3 epitope disappears.

Through mutational analyses and '*domain-swapping*' experiments in the laboratory of Irving Boime in St. Louis, Missouri, and in other laboratories, it became indisputably clear that the ECD represents *the* ligand-binding domain, and that the glycane residue bound at the asparagine-52 of the α-subunit of hCG is essential for hCG's agonist-potency.

In 1993, Kobe and Deisenhofer (Houston, Tex., USA) succeeded in clarifying the **crystal structure** of the **RNase1/angiogenin inhibitor** (RAI), which, quite interestingly, is formed of 16 leucine-rich repeats, each of them similar in length and sequence to those of the ECD of the hCG receptor. Surprisingly, RAI turned out to have a non-globular structure in which the ensemble of the 16 repeat (each a block or '*module*' being an '*independent folding domain*') forms a horseshoe-shaped structure. Based on these findings, we revised once more our picture of the hCG receptor of 1988 (see picture above right).

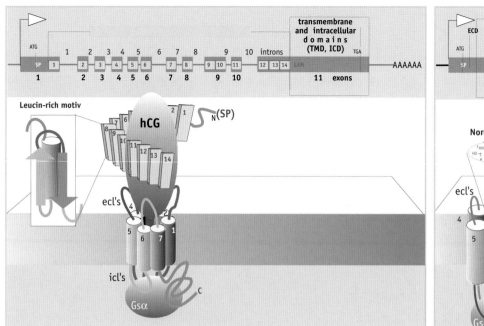
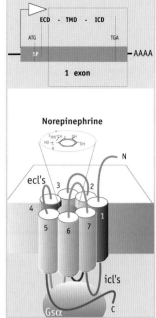

hCG receptor | Adrenergic receptor

The **gene structure of the hCG receptor** was elucidated in the laboratory of Maria DUFAU at the National Institutes of Health (NIH) in Bethesda, Maryland in 1991. As the picture above shows, this gene consists of 11 exons, with exon 11 encoding the entire transmembrane block, including the extra- and intracellular loops (ecl, icl) and the intracellular domain (ICD), and exons 1 to 10 each encode one leucine-rich repeat of the ECD. Exceptions are repeats 9 and 10, which are coded for by exon 9, and repeats 12, 13 and 14 which, together with the transmembrane block, are encoded by exon 11.

Thus, the hCG receptor differs from all other members of the 7TMHR superfamily, which are almost exclusively encoded by a single exon and are without introns (see picture on the right): The hCGR displays a *mosaic gene structure*: one part of this evolved by gene enlargement via several *gene duplication events* of a small exon encoding a *single ancestral leucine-rich module* to produce the ECD, and the other part, i.e. the *ancestral single exon 7TMHR gene*, through *exon shuffling*, became linked to the former multiexon gene.

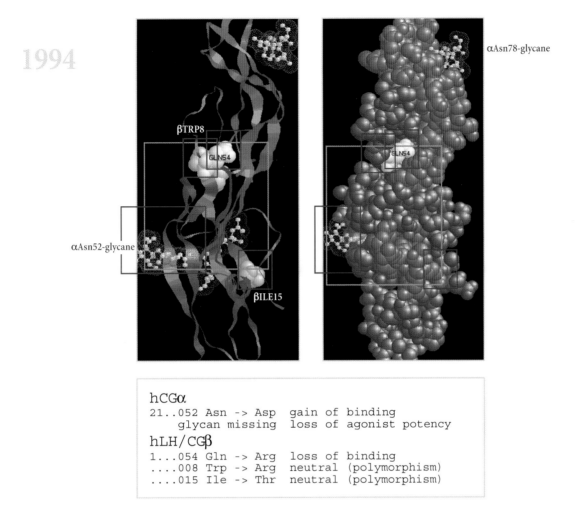

```
hCGα
21..052 Asn -> Asp   gain of binding
       glycan missing   loss of agonist potency
hLH/CGβ
1...054 Gln -> Arg   loss of binding
....008 Trp -> Arg   neutral (polymorphism)
....015 Ile -> Thr   neutral (polymorphism)
```

Elucidation of the structure of a protein, no matter by which method, is prerequisite for understanding its function and the manifold mechanisms therein. A structural explanation provides, among others, knowledge that a protein is constituted of *functionally distinct domains*, each endowed with a different task.

Deliberately introduced mutations as well as those that occur naturally and the more or less pathological *phenotypes* resulting from them, permits establishment of a causal link between structure and function to elucidate which domain performs which function.

In 1994, A. J. Lapthorn (laboratory of N.W. Isaacs, Glasgow) succeeded in clarifying the **crystal structure of hCG**. They demonstrated that the biologically-active α:β-heterodimer hCG is an ovoid globular protein (for further details see chapter on 'Peptide Hormones'). The white frames in the pictures above show the rough borders of the receptor interaction surface that come into contact with the β-sheet structures inside the horseshoe-shaped ECD of the hCG receptor. It can be clearly seen that parts of both subunits are involved and that the αAsn52-glycane moiety is also included. Thus, it is confirmed that αAsn52-deglycosylated hCG can bind, but no longer activate, the hCG receptor, thus being a competitive antagonist.

In 1992, for the first time, a **mutation in the lutenizing hormone LHβ gene** (which is quite similar to hCGβ) was described that leads to an *immunologically intact*, but *biologically inactive*, LH, and thus to a female phenotype with male karyotype (46, XY). In this mutated LHβ, glutamine (Gln)-54 is replaced by an arginine. As shown above, this Gln is involved in the *dimerization* with the α-subunit and is also part of the *receptor contact surface*. The consequence of this mutation is that the *competence* for both heterodimerization and receptor binding is diminished or lost.

Other mutations in LHβ, such as Trp8 and Ile15, were also described, but turned out to be neutral. Why? Because, as the pictures above also show, they are located in the back of the LH/hCG (therefore hidden in the picture to the right), i.e. that part that does not participate in receptor interaction.

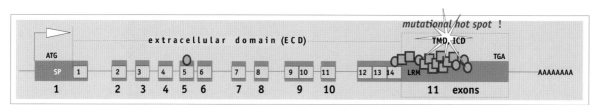

From 1995 onwards, numerous **mutations in the hCG receptor gene** were described and deposited in the OMIM (Online Mendelian Inheritance in Men Data Bank). We have compiled them on a *gene mutation map* (as depicted above), from which it can be seen that the majority of the mutations occur in exon 11, although from a statistical point of view, the entire gene should be equally susceptible to mutations. A gene region with so many more mutations than expected is called a *hot spot*. There are a variety of reasons for such an accumulation of mutations but, to a large extent, these remain to be elucidated.

Nevertheless, the effect of a mutation of the gene on the respective protein's function can be readily seen if the location of the mutation is mapped into the protein's structure. Exon 11 encodes the entire transmembrane block and is involved with signaling to the cell's interior (transmitting the signal's message from outside to the inside of the target cell) rather than with ligand binding. A mutation in exon 5 = 5th leucine repeat of the ECD, has the expected *loss of binding* as a consequence (i.e., a reduction in ligand binding affinity). The phenotype of a male patient afflicted with this mutation is described to have a microphallus and a reduced number and function of testosterone-producing Leydig cells. They display reduced hCG binding (affinity/capacity), but a residual ability to generate cAMP, so that a male phenotype is still guaranteed. Theoretically, there can also be a gain of binding through a different mutation in the ECD, but such has so far not been described.

Other loss of function mutations have been found, all of which are located in the hot-spot region mentioned above. As can be seen in the next picture, the allelic variants 8 and 9 are *missense mutations* that lead to a premature stop codon and thus to a truncated receptor. This can still be embedded in the membrane, but cannot couple with the G-protein, since the icl-3, the 6th and 7th TMH and the ICD are missing. Allelic variants 19 and 20 (in 7th TMH) also lead to a loss of signaling ability, presumably because this helix becomes inappropriately positioned, as does icl-3. Mutation 19 appears less severe than mutation 20, possibly because it is less involved in coupling.

All other loss of signaling mutations lead to a 'severe phenotype', i.e. male pseudohermaphroditism (karyotype 46, XY) with complete feminization of external sexual characteristics, creating individuals who are, from social, psychological and legal points of view, regarded as females. Since the development of a female embryo is not dependent on the presence and functioning of LH or hCG or the LH/hCG receptor, respectively, individuals with a female genotype (46, XX) exhibit a milder phenotype that, at most, manifests itself as sterility in adult life.

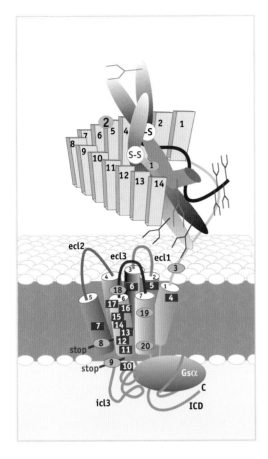

```
hCGR
 2...133 Arg -> Cys        loss of binding
 3...354 Glu -> Lys        loss of signaling
 4...373 Ala -> Val        gain of signaling
 5...398 Met -> Thr        gain of signaling severe
 6...457 Leu -> arg        gain of signaling
 7...542 Ile -> Leu        gain of signaling
 8...545 Cys -> Stop       loss of signaling
 9...554 Arg -> Stop       loss of signaling
10..564 Asp -> Gly         gain of signaling
11..568 Ala -> Val         gain of signaling
12..571 Met -> Ile         gain of signaling
13..572 Ala -> Val         gain of signaling
14..577 Thr -> Ile         gain of signaling
15..578 Asp -> Gly         gain of signaling mild
16..578 Asp -> Tyr         gain of signaling
17..581 Cys -> Arg         gain of signaling
18..593 Ala -> Pro         loss of signaling
19..616 Ser -> Tyr         loss of signaling mild
20..625 Ile -> Lys         loss of signaling
```

Mutation 'map' of the hCG receptor. The numbers in the inset represent a total of 19 allelic variants protocolled in the OMIM data bank to date. Next to them, the corresponding AA sequence number is given together with the mutated AA or a stop codon, respectively. Furthermore, the alteration in function of the receptor protein consequent to the respective mutation is indicated: Loss or gain of function. Both altered functions can be further subdivided into alterations in *binding* or *signaling* capacity. The locations of these 19 mutations are shown both in the gene map (without numbers) and in the protein model [with numbers and symbols that differentiate between loss (ellipse) and gain of function (square)]. (Mutations #1 and #21 refer to those in LHβ and hCGα genes and are shown on page xxiv).

Loss of function mutations have long been known for numerous proteins, and their resulting dysfunctions on the physiological level are quite well understood and have even been predicted depending on the protein concerned: an enzyme deficiency is the most generally used example. It was also clear that, given adequate biological redundancy, the expression of half the amount of a functioning protein should be sufficient to prevent occurrence of a pathological phenotype in the *heterozygotic* condition (characteristic of recessiveness).

However, the understanding at the molecular level that mutations can also result in a *gain of function* has been achieved only in the past few years. In the case of a receptor, such mutations could yield an increased affinity for the corresponding agonist, or, affinity and occupancy being unchanged, could result in more potent receptor coupling that, in the most severe cases, could take place even in the complete absence of an agonist. Such a mutation would represent a *constitutively active receptor* that couples constantly and indefinitely.

The consequences in the case of a growth factor receptor would be, quite as expected, excessive proliferation ending in tumor growth (such a receptor being one of the many *oncogenes*). In the case of the hCG receptor, Leydig cell hyperplasia with increased secretion of testosterone resulting in *familial male precocious puberty*, also called testotoxicosis, is the result.

A constitutively unstoppably-active receptor manifests itself by increased cellular signaling even when the other allele codes for the normal, agonist-dependent receptor whose activity is regulated. Under these conditions, even in the *heterozygotic* condition, a pathological phenotype occurs, and this called a *dominant* mutation.

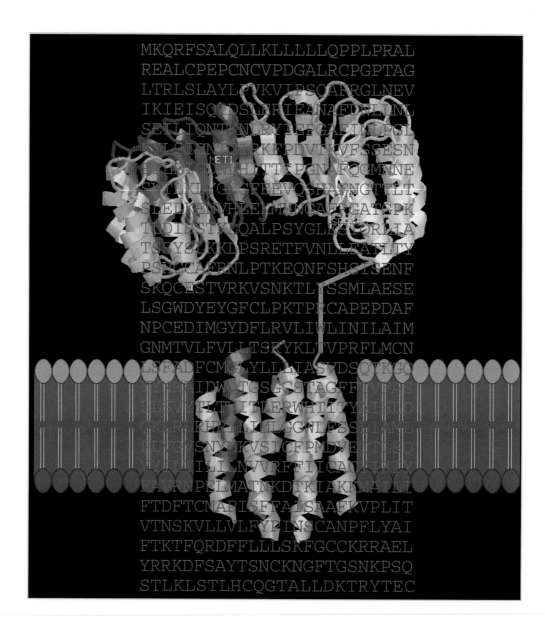

This picture shows the 'putative' 3-dimensional structure of the hCG receptor as well as its topology within the plasma membrane. Superimposed is the primary structure (AA sequence). The receptor is formed of a transmembrane block of 7 α-helices (corresponding to the *visual receptor rhodopsin*, the loops and the intracellular domains not being modeled) and the ECD, displaying the horseshoe-like structure similar to RAI, as already mentioned. The interior surface of the ECD most probably represents the ligand contact surface. As will be shown in the chapter on 'Membrane Receptors', the space inside this horseshoe is both macroscopically large enough and microscopically suitable for accommodating hCG and for specifically binding it, respectively. This then induces a productive conformational transmission through the transmembrane block towards the cytosolic surface of the receptor that, in turn, allows its *coupling domain* to couple and thus to activate the stimulatory GTP binding protein Gsα.

Ligand-Binding Proteins – Large Molecules

Enzymes

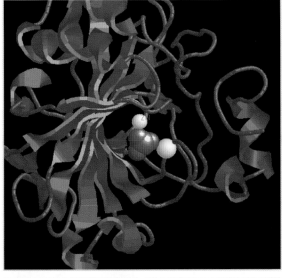

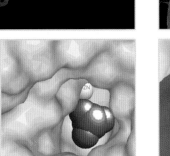

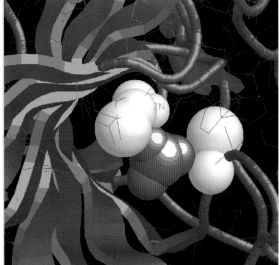

Carbonic anhydrase

(PDB 1CAM)

Enzymes catalyze chemical reactions that otherwise, due to their *thermodynamics*, are unlikely. Enzymes bridge these *improbabilities* simply by bringing the reactants closer together. Enzymes possess this capacity because, as proteins, they fundamentally have an endless variety of forms, which is why they can bind an endless variety of *substrates*, i.e. molecules to be chemically modified. For this purpose, enzymes possess a *substrate-binding domain*. The basis of every enzymatic reaction is this proper (i.e. *with atomic precision!*) positioning of a substrate adjacent to a certain reactive group within the enzyme itself. This fundamental property (i.e. infinite number of binding specificities) is also shared by other macromolecules that are actually not enzymes, such as antibodies or RNAs, but can occasionally also act as effective catalysts!

After enzyme binding, the substrate often acts like an allosteric activator in inducing a certain conformational change of the enzyme (of course, to a lesser degree than an agonist would induce on a receptor). Consequent to this or the transient binding itself, the substrate comes into immediate proximity to reactive groups within the enzyme, i.e. to specific amino acid (AA) residues in the *active center* (often identical with the substrate-binding site) plus other possible *co-factors*, which are bound in a separate pocket but are still in spatial proximity to the substrate, e.g. ATP in the ATP-binding pocket of a kinase. If all the involved reactants are positioned adequately close to each other, then a 'switch' can occur, i.e. a transfer reaction (e.g. a phosphate residue from an enzyme-bound ATP can be transferred to a protein to be phosphorylated) or, conversely, a removal reaction of a specific group (e.g. dephosphorylation or demethylation etc.), or an isomeric rearrangement (e.g. a Δ4 steroid converted to a Δ5 steroid), a ligation or a hydrolytic cleavage to, or of, larger polymeric substrates (e.g. proteins, RNA, DNA), respectively, among other reactions. All this is made possible by transient chemical alterations or steric rearrangements of certain AA residues within the active center of the enzyme (including, in certain cases, some AAs outside of that active center).

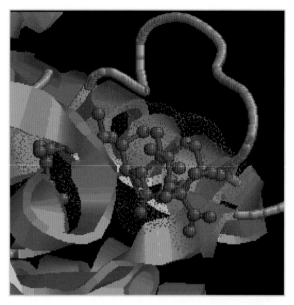 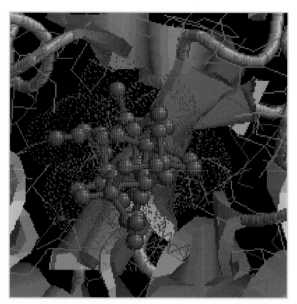

3α, 20β-HO-Steroid dehydrogenase + steroidal inhibitor (Carbenoxine) (PDB 1HDC)

The International Enzyme Commission has classified enzymes as (1) oxidoreductases; (2) transferases; (3) hydrolases; (4) lyases; (5) isomerases, and (6) ligases.

Catalytically supported reactions mostly follow a certain sequence of steps, which implies that the enzyme undergoes *consecutive conformational changes*, especially in the region of its active center. Many enzymes require the cooperation of coenzymes, that, not being proteins themselves, must contact the enzyme via an appropriate interaction site. Such contact sites need not be permanently accessible, i.e., they can be masked, but made transiently accessible by allosterically induced conformational changes.

Every enzyme must be localized within the cell at its correct place, a prerequisite that is met by *targeting domains* (or *signals* or *sequences* or *motifs*) on the enzyme itself. This possession of *destination flags* applies, of course, to all proteins in general! One of the discoverers of this principle, Günter BLOBEL (Rockefeller University, New York), was awarded the Nobel prize in Medicine in 1999.

Enzymes are the essential molecules of life because, in a certain sense, they 'act against' thermodynamics, *as life itself* does (actually, enzymes just facilitate and speed up reactions that are otherwise extremely slow). Enzymes are involved in the 'usual' processes of intermediary metabolism, e.g., transforming nutrients such as glucose into building blocks such as acetyl-CoA (coenzyme A) that, in turn, are required for synthesizing other molecules, e.g., steroid hormones, or in the larger anabolic processes, such as the synthesis of molecular polymers (proteins, RNA, DNA).

Enzymes are also constituents of the different *signaling cascades*, e.g. receptors for growth factors can possess an intracellular domain that has catalytic properties such as a protein-tyrosine kinase. 7TMHR-associated G proteins have an intrinsic GTPase activity. The link between signal transduction and the effector machinery of a cell is primarily provided by cytosolic enzymes, namely protein kinases. Moreover, some hormones can be multifunctional and thus also act as enzyme inhibitors (e.g. angiogenin as an RNase inhibitor, RAI). Viruses could not be replicated by the host cells if they did not themselves bear and bring along the genes for certain enzymes. The proteases of HIV are examples of enzymes that split a viral multiprotein precursor into individual active proteins. Reverse transcriptase is essential for the transcription of the viral RNA-genome into host-incorporable DNA, a prerequisite for replication of RNA or retroviruses, e.g. HIV.

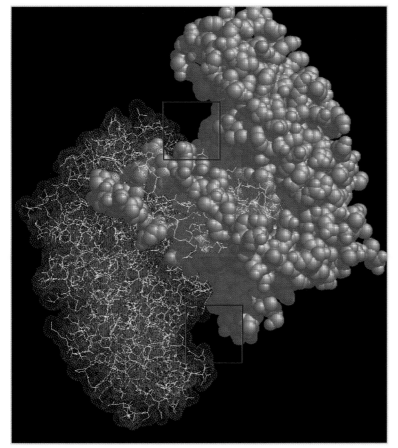

Citric acid synthetase dimer (PDB 5CSC)

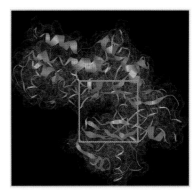

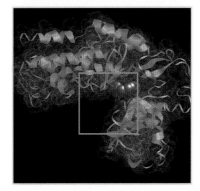

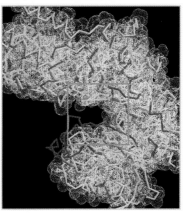

Note the conformational change in the center of the **hexokinase** upon substrate binding! The two lobes of the enzyme come closer to each other and thereby tightly bind the substrate: *induced fit*! The lower picture shows the two enzyme states (white: empty, magenta: filled/liganded) superimposed over each other.

Hexokinase ± glucose (PDB 1HKG, 2YHX)

All proteins that must bind ligands require the presence of a *ligand-binding pocket* configured to correspond in size, structure and physicochemical complementarity to the ligand to be bound. Note in the left picture how both monomers of citric acid synthetase penetrate each other! A similar situation has previously been shown in α:β holo-hCG. The enzyme-dimer depicted here is seen in the *open conformation*, as indicated by the shape of both *grooves* (in frames). There is, of course, also a *closed* conformation, as exemplified with another enzyme shown here, i.e. hexokinase ± glucose.

If an AA in this pocket is mutated, or outside of it, but exerting a conformation-disturbing effect on the pocket, the ligand can bind only with lower affinity or not at all (sometimes, however, it can also be bound even better!). In principle, the same is true for mutations outside the pocket that may exert far-reaching conformation-disturbing effects on the ligand-binding pocket. In any case, because the enzyme is less active or inactive, the corresponding metabolic or signal transduction step cannot occur. So the substrate (or ligand molecule in general) accumulates and, now present in higher than normal concentrations, can make itself 'noticeable' by interactions with other proteins with which it would and should not interact at normal concentrations and under normal circumstances.

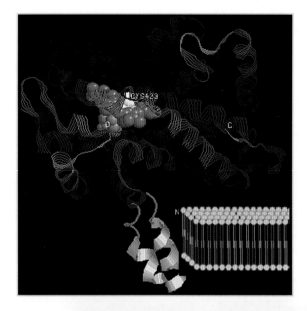
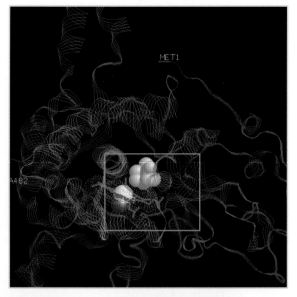
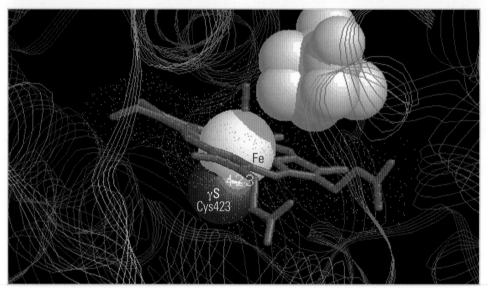

P450 Side chain cleavage enzyme + Hem (PDB 1SCC)

Successful survey of the 3-D shape of such a pocket should permit construction of a ligand that fits with even higher affinity than the natural ligand, preventing the latter from being bound and catalytically altered. The enzyme would thus become blocked, although the result would, at least transiently, be the same as if the enzyme was structurally disturbed due to a gene mutation. Like all biomolecules that are effective only through their interaction with cellular target molecules, certain drugs are of therapeutic or diagnostic relevance since they act as inhibitors of specific enzymes.

Angiotensin-converting enzyme (ACE) inhibitors are among the most important drugs against high blood pressure. *Renin inhibitors* can also be used for this purpose. *Protease inhibitors* are, as previously mentioned, essential for blocking the multiplication of HIV and thus the manifestation of AIDS, but alone are not sufficient, requiring administration of further inhibitors, e.g. inhibitors of the *HIV reverse transcriptase*. *Metopiron*, an inhibitor of steroid-11β-hydroxylase, is used for diagnostic purposes, to prove or disprove a suspected pituitary insufficiency. *Recombinant angiogenin*, an inhibitor of several RNases, is applied to cellular RNA extracts to protect RNA against degradation by endogenous RNases. Isolation of intact mRNA is a technical prerequisite for diagnostic detection of mutations.

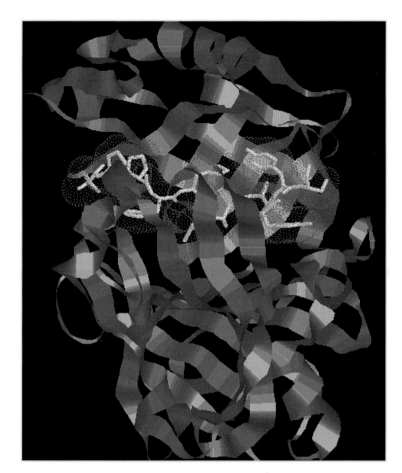
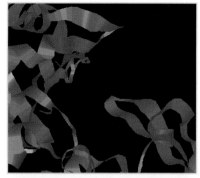
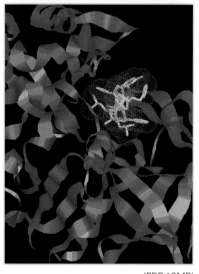

Renin ± peptidic inhibitor in different views. The inhibitor represents the sequence His6-Pro-Phe-His-Leu-Leu-Tyr-Tyr-Ser14 of angiotensinogen. Note the similarity of this human enzyme with the viral HIV protease (see following page).

(PDB 1SMR)

The pictures in this chapter show examples of pockets, from the smallest to the largest, within proteins. As previously mentioned, the substrate-binding pocket is often the active center of an enzyme which can possess a further *subsite*, or a 'niche', for a specific co-factor. As intermediary carriers of atoms or groups of atoms, such co-factors (also called *coenzymes* or *prosthetic groups*) are essential. Co-factors can also vary in size (as do the respective subsites). It can be a nucleoside, such as ATP or GTP and other compounds of similar size. A single atom or ion, e.g. a Ca, Mg, Mn, Mo, Se, Cu, Zn or Fe, held by certain AAs covalently or in ion binding, respectively, can cause a certain labile domain to be correctly oriented in the entire architecture of a multidomain protein. While most ions are always coordinated by appropriate AAs (as previously described for calcium binding, or as will be described later for zinc binding between γ-sulphur atoms of cysteines in transcription factors), an iron atom is generally complexed within a much larger non-AA molecule, e.g. a porphyrine ring (the hem group in members of the cytochrome P450 enzyme superfamily or in hemoglobin and myoglobin), rather than exclusively between AAs (comparable to Ca^{2+} binding to A23187, page 15).

The picture in the left upper corner on page 34 shows two α-helices as parts of the enzyme. These can be considered as targeting sequences (as alluded to above) that anchor the enzyme in the inner mitochondrial membrane. This lipophilic mitochondrial domain also serves to accumulate and partition lipophilic substrates from the aqueous environment of the cytosol and to allow them a kind of vectorial entry into the substrate binding pocket containing the hem group.

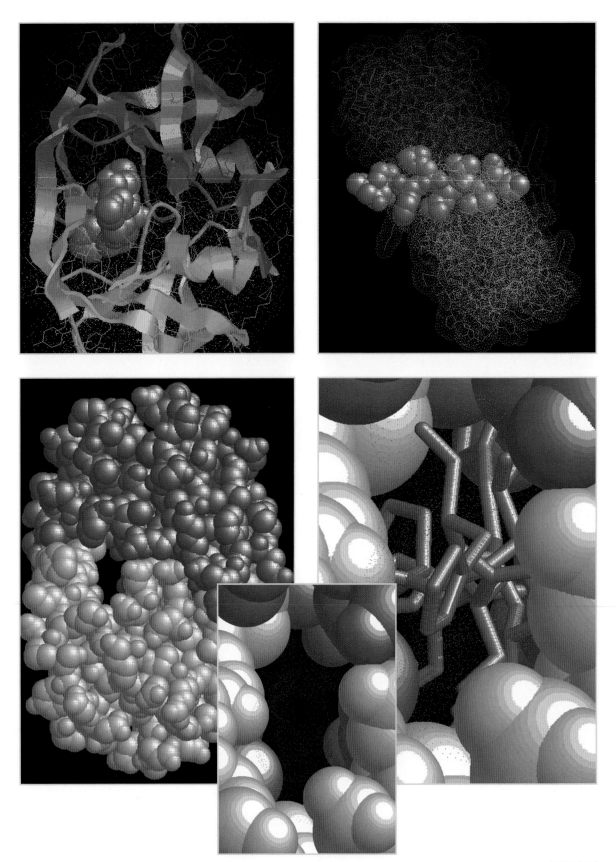

HIV2 protease ± renin inhibitor (His-Pro-Phe-Leu-Val-Ile-His) in different views (PDB 2PHV)

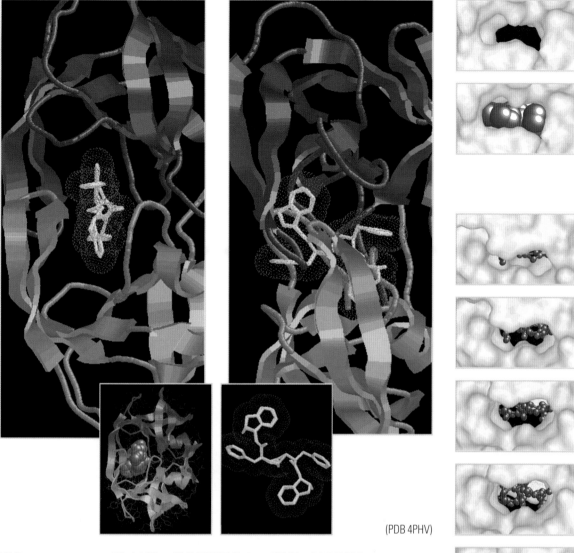

(PDB 4PHV)

HIV1 protease + nonpeptidic inhibitor: N, N-BIS(2(R)-Hydroxy-1(S)-idanyl-2,6-(R,R)-3-diphenylmethyl-4-hydroxy-1,7-heptandiamide. Left lower inset: **HIV2 protease + peptidic inhibitor** (PDB 2PHV). Note the similarity of this viral enzyme with human renin!

Enzyme inhibitors can have peptidic structure. Often they contain atypical AA residues, e.g. a derivatized (e.g. methylated) AA, or a *D-* instead of the natural *L-*amino acid. The peptide may also be cyclicized and thereby made conformationally rigid (in contrast to the natural ligand that may be a linear peptide with abundant conformational freedom), among other modifications.

Inhibitors can also have a nonpeptidic structure and a given enzyme can be blocked by peptidic as well as nonpeptidic inhibitors. Inhibitors can occupy the natural substrate-binding site, but may also occupy a *side pocket* within it, or fill yet another, *ectopic* site of the enzyme. It is important that they inhibit the enzyme's catalytic activity, either direct-competitively or indirect-allosterically. Since many enzymes can act only as dimers, some drugs can exert their inhibitory effect also by occupying the protein-protein interaction domains (dimerization contact sites) without themselves being inhibitory for the catalytic activity of the enzyme.

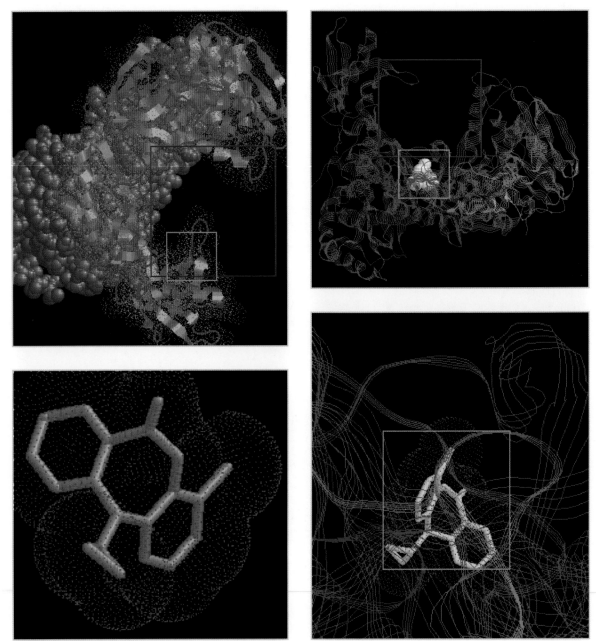

HIV1 reverse transcriptase homodimer + **dipyrodiazepinone** (Nevirapin)-inhibitor. (PDB 3HVT)
The frame indicates the RNA/DNA *binding cleft* (active center). The inhibitor is situated in a subsite of the active center. Note that Nevirapin is a non-nucleoside molecule and thus leaves the human body's own nucleotide transferases largely undisturbed.

All that has been said above about enzyme inhibitors also applies to receptor ligands: think of morphine (the alkaloid) and the endogenous opioids, which are peptides, or of tranquilizers of the benzodiazapem class, which bind at a site other than the agonist-binding site to the GABA-gated chloride channel, and yet can activate/open it just as GABA, albeit in a qualitatively different manner.

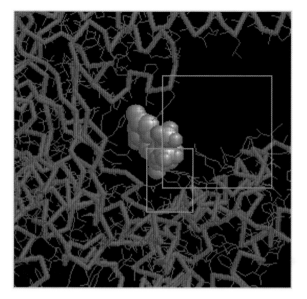 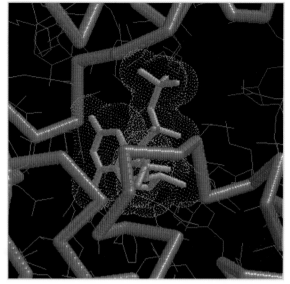

Human DNA polymerase (KLENOW-fragment) + dCTP in different views (PDB 1KFD)

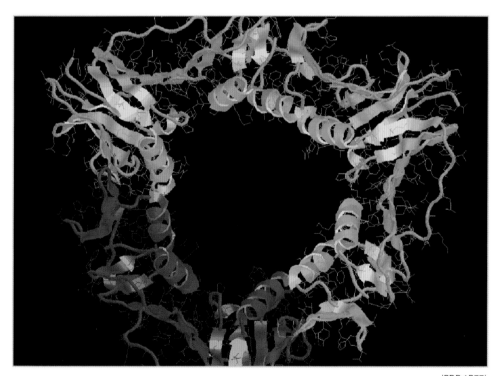

DNA sliding clamp of bacteriophage R869 DNA polymerase (PDB 1B77)

The larger the substrate, the larger the substrate-binding site has to be. Here, in DNA polymerase, it is not a *pocket*, it is a *cleft* in which DNA, the substrate, binds. A single nucleotide to be transferred, however, such as the dCTP shown here, is situated in a small subsite of the large substrate-binding *cleft*.

In the *sliding clamp* subunit of DNA polymerase, it is not a *cleft*, it is a *large channel* in which the template DNA together with newly synthesized DNA become accommodated.

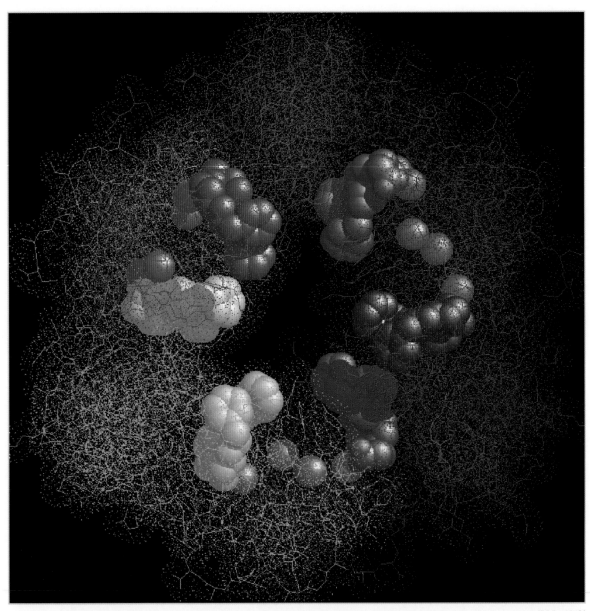

Secretory phospholipase A2-hexamer, as found in inflammatory synovial fluid. In this picture, each monomer is complexed with 4 Ca^{2+} ions plus one molecule of an inhibitor: 4-(S)-[(1-Oxo-7-phenylheptyl)amino]-5-[4-(phenylmethyl)phenylthio]pentanolic acid (OAP).

(PDB 1KVO)

The picture above shows a highly complex protein aggregate, a hexamer of the enzyme phospholipase A2. Phospholipase A2 species occur intracellularly, where they hydrolyze arachidonic acid from cell membrane phospholipids, serving then as substrate for the synthesis of prostaglandins. Furthermore, Ca^{2+}-dependent phospholipases can degrade membrane phospholipids to such an extent that cell destruction results. Phospholipase A2 can, however, also be secreted, among others, by pancreatic acinar cells, to split dietary phospholipids, or by inflammatory cells, that thus produce tissue damage.

Knowing this allows one to envisage a novel pharmacological approach for an anti-inflammatory therapy, i.e. by selective phospholipase A2 inhibitors.

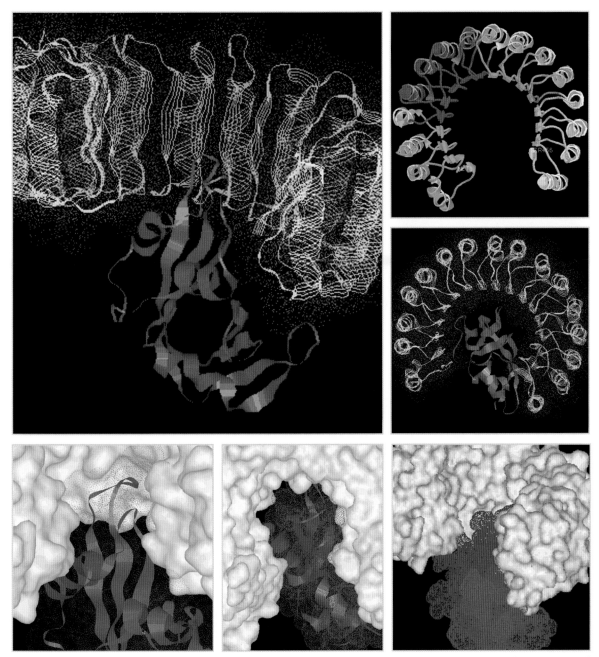

RNase inhibitor (white) ± angiogenin (magenta) in different views (PDB 1BNH, 1A4Y)

As previously stated, the substrate binding site must be commensurate to the size of the substrate. The same can be true for an enzyme inhibitor. In the example shown here, it is no longer a pocket or a cleft, but literally a *deep cave*! It is clear that for an enzyme of the size of angiogenin, the inhibitor has to be quite large: RNase/Angiogenin Inhibitor (RAI). As can be seen, this inhibitor engulfs the enzyme and thus prevents substrate (i.e. RNA) binding.

As one can see again: Enzyme inhibitors need not always be chemically the same as the orthologous substrate. A protein can compete, displace or sterically hinder RNA, a nonpeptidic ligand can prevent proteolytic activity (see HIV protease!), a non-nucleoside analog can block DNA synthesis (see HIV reverse transcriptase), and there are numerous such examples.

It is the *fit*, not the chemical class that is important, which is also true, as previously mentioned, for receptor ligands and any other pair of ligand and ligand binding molecules.

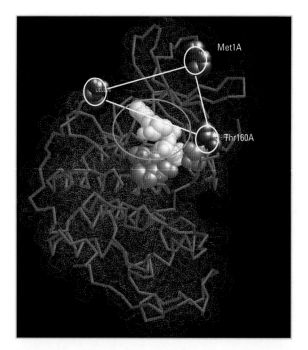 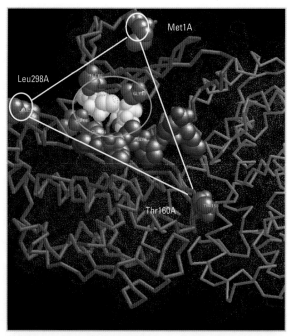

CdK2 (magenta) without cyclin A (left panel) and complexed with **cyclin A** (grey, right panel); white CPKs: **ATP** within ATP-binding pocket of CdK2; magenta CPKs: AA residues lining the entrance of the ATP-binding pocket; dark magenta CPKs: N and C terminal AA residues (Met1 and Thr160, respectively) and Leu 298.

(PDB 1B38, PDB 1QMZ)

As has been pointed out previously several times, induction of a conformational change in a protein is crucial for modulating its activity, either activating it, when it is in an inactive state, or vice versa. Modulation can be effected by small agonists ('ligands') that induce an allosteric change or by large proteins that attach to the target protein over a broader area (referred to as *protein-protein interaction*). An example of the latter possibility is shown on this page with an enzyme most critical for cell cycle control, i.e. CdK2 (cyclin-dependent kinase 2). The left panel shows the enzyme in its cyclin A-uncomplexed, and hence inactive, state. Note the size and corners of the 'landmarks triangle'! As can be seen on the right panel, the complex of CdK2 with cyclin A (to the right side) leads to a major conformational change (enlargement of the triangle!) and opening of the substrate-binding pocket, a subsite of which harbors ATP (white CPKs). ATP is bound already in the inactive state, but can not be hydrolyzed and utilized for transphosphorylation, as no substrate protein can be bound. Below the ATP pocket one can see a substrate peptide (residues 1–7, dark grey CPKs) from a substrate protein bound to the active CdK2.

Membrane Receptors

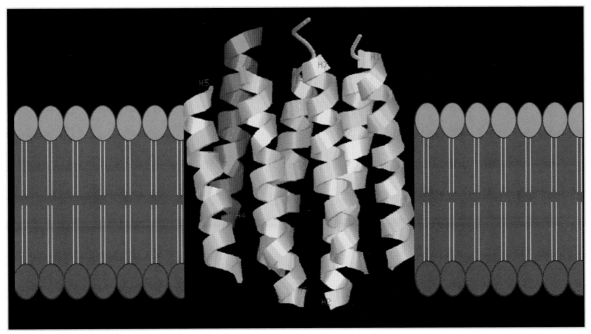

Rhodopsin (a 7-transmembrane helix receptor) without retinal (side view of the seven transmembrane helices) (PDB 1BRD)

Receptor proteins are the signal-receiving molecules of a cell. Signaling is the means of intercellular communication, as introduced at the beginning of this book. Even a single-cell organism cannot survive without receiving signals from the 'outside world': An amoeba would never find food were it not directed there by food chemotaxis receptors; a rhodobacter bacterium would never survive had it not a receptor for light.

How much more is this dependence on communication true for the life of a multicellular organism in which exchange of signals between organs, cells, and even between organelles inside the cell is of vital importance, not to mention the reception of signals from outside, of light, olfactory and gustatory signals, food substances, antigens, toxins, drugs etc. For all of these internal as well as external signals there are receptors. Interestingly, receptors for exogenous signals are not constructed much differently than those for the endogenous signals! What matters is the physicochemical nature of the signals, not their origin: hydrophilic (polar) vs. hydrophobic (apolar, lipophilic). The latter can freely diffuse across the cell membrane and so find their cell-internal receptors. These are actually almost exclusively transcription factors and will be discussed in detail later (i.e. in chapter 'DNA-Binding Proteins').

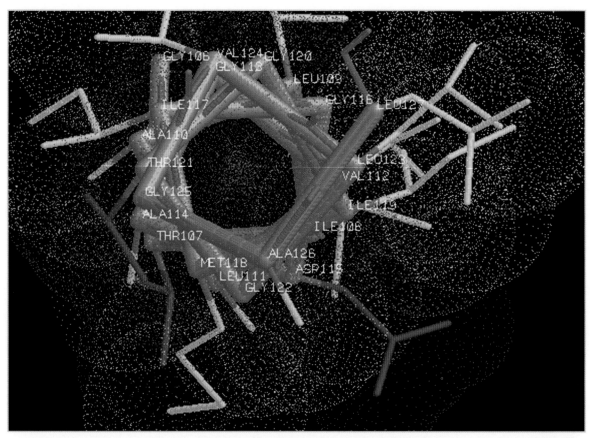

Rhodopsin. The 4th helix seen axially from above, lipophilic amino acids white, hydrophilic magenta.

(PDB 1BRD)

Hydrophilic compounds in general, and hydrophilic signals in particular, cannot pass the cell membrane because they are not soluble in a lipophilic fluid or semifluid matrix. Therefore, the plasma membrane must contain receptors for hydrophilic signals. These are proteins of complex structure and of a multitude of domains.

Basically, a ligand-binding domain faces the exterior of a cell, while the coupling or effector domain faces the interior or cytosolic side. In between these two, embedded in the membrane, is the *membrane-anchoring domain*. As mentioned in the context of enzymes, the extracellular signal binds to the *ligand-binding domain* and thus acts as an allosteric activator of the receptor. This means, the bound signal induces a conformational change in the receptor that is transmitted through the membrane-anchoring domain onto the *coupling domain* localized at the inner face of the cell wall. Although the signal itself has physically not reached the cell's interior, the *signal's message, by being converted to receptor form*, leads to a cascade of intracellular biochemical consequences as if it had really entered the cell. This begins with the receptor's coupling domain, now rendered competent to interact with intracellular proteins that, thus activated, can interact with further proteins, and so on, finally leading to a *cellular response* to that extracellular signal.

The membrane-anchoring domain of any membrane protein, and thus of membrane receptors as well, generally consists of one or more α-helices that have 20–25 amino acids (AAs), mostly lipophilic: Ala (A), Val (V), Ile (I), Leu (L), Gly (G), Phe (F). Why 20–25? It is an intrinsic property of α-helices that the carbon-backbone of a peptide string made of helix-forming AAs is bent both in the X as well as the Y axes so that a circular shape in the X plane and and a screw-shaped descending spiral in the Y plane results, with 3 and 4 AAs alternately making turns. Five to six turns (20–25 AAs) provide for an α-helix that is long enough to vertically traverse the lipid bilayer cell membrane, which is about 160 Å thick. The AA residues point vertically away from the the helix's longitudinal axis and thus build a large lipophilic interaction surface that can readily dissolve in the lipid membrane (compare insert page xxi).

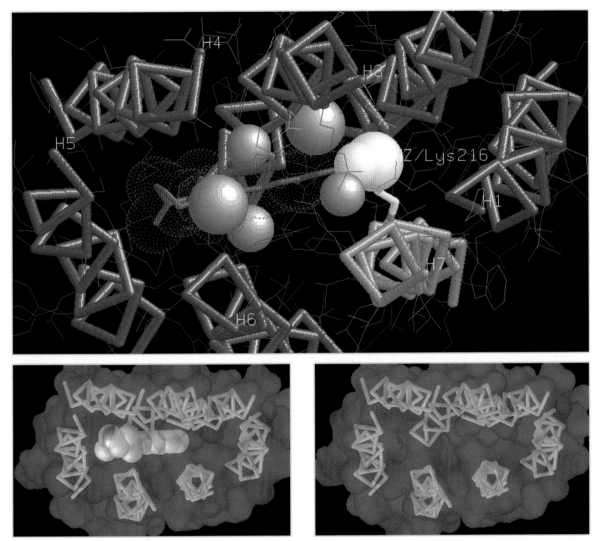

Rhodopsin + *all-trans*-retinal. Axially seen from above, retinal-contacting atoms of rhodopsin shown as CPKs. (PDB 1BRD)

However, α-helices also can be amphipathic, i.e. built of both lipophilic and hydrophilic AAs. As the axial view of the 4th helix of rhodopsin on the previous page shows, the lipophilic (light grey) AAs localize on one side (top) and the hydrophilic AAs (magenta) on the other (bottom) so that the helix as a whole has two physicochemically different faces, i.e. is amphiphilic. Several such amphiphilic helices can build a large palisade-like ensemble, called 7TMHR, where 7 helices form an *ellipsoid helix* bundle in which all their lipophilic surfaces point outwards to the membrane and the hydrophilic surfaces point inwards, i.e. towards the room surrounded by the palisade. This inner space represents a vertical elipsoid pocket with a diameter of approx. 6x12 Å into which ligands of appropriate size can *fly* from outside. This basic architecture of 7TMHRs is suitable for all the small ligands that were presented at the beginning of this book: catecholamines, morphines/opiates and most other neurotransmitters and neuropeptides.

This receptor construction is suitable even for receiving nonmaterial (nonmolecular) physical signals, namely light! Since, however, every sort of signal eventually has to be transformed into a chemical signal, evolution has resorted to a trick in placing a permanent ligand into the light receptor, where it is bound covalently. Rhodoposin is the photon receptor and 11-*cis*-retinal is the photon-sensitive ligand that, in the form of a Schiff's base, is bound to lysine 216 of the 7th TMH of the receptor, which is a hydrophilic AA and thus points towards the pocket's interior. Through additional, but non-covalent, interactions with other AAs protruding from vis-à-vis located helices, the retinal acquires an almost horizontal position. In contrast to most of the small molecules, the 11-*cis*-retinal is not rigid, but flexible, i.e. it can be bent at a specific point in response to light.

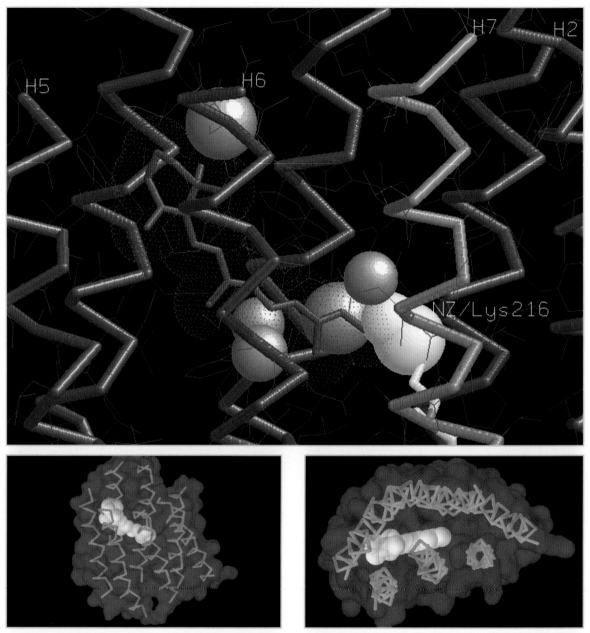

Rhodopsin + *all-trans*-Retinal. Side view, retinal-contacting atoms of rhodopsin shown as CPKs.   (PDB 1BRD)

Light induces isomerization of 11-*cis*-retinal to *all-trans*-retinal, that is, rearrangement of the binding angle between C12 and C13. The *all-trans*-retinal as a whole is thus raised by a few degrees from its horizontal position in the pocket (as shown in the picture). Change of position of a ligand must, just as the *binding* of a ligand, induce a conformational change in the receptor protein, which is what is really at stake. Thus, an 'immaterial', nonchemical signal can also be converted into a chemical signal (receptor activation).

There are many in favor of the idea that, early in evolution, there was an archetypical 20-25 amino acid-long transmembrane helix (TMH) module gene that, by duplication and modification, multiplied and gave rise to thousands of different combinations of membrane proteins. Thus, in the course of evolution, and quite early, given its existence in bacteria, membrane proteins with 1, 3, 4, 7, 10, 12, 24, etc. TMHs evolved and founded their own superfamilies and families. These membrane proteins acquired different functions from their respective differing structural organizations. *Receptors*, *cell adhesion proteins*, *transporters* (members of the *solute carrier superfamily*, '*uptaker*'), *ion pumps* (ATPases) and *ion channels*, to mention the most important ones.

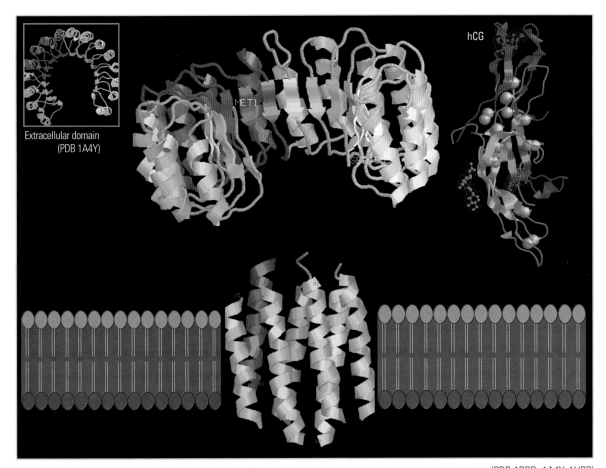

Putative **HCG** receptor ± hCG (PDB 1BRD, 1A4Y, 1HRP)

The superfamily of the 7TMHR comprises a large group within the receptor class of membrane proteins. Small ligands, such as norepinephrine or morphine, are bound within the 7TMH pocket, while larger ones, such as peptides and proteohormones, would not fit. Therefore, evolution has created, by *exon shuffling*, not only multihelix proteins, but also proteins that represent mosaics of stones taken from different gene families. As a matter of fact, for large ligands, 7TMHR evolved long and complex extracellular ligand-binding domains, as exemplified here with the hCG receptor. The ligand-binding domain itself consists of small (about 25 amino acid-long) *leucine-rich repeats*, each built of a β-sheet and an α-helix. Such a building block obviously can be put together into any multiples to create a larger molecular entity. The extracellular domain of the hCG receptor possesses 14 such modules, similar to another protein, RNase/angiogenin inhibitor (RAI), described earlier which possesses 16. Together, these build a kind of horseshoe-shaped structure that forms an inner space that is both large enough and variable in detailed shapes so that very different large proteins, such as RNase in one case and hCG in the other case, can be accommodated.

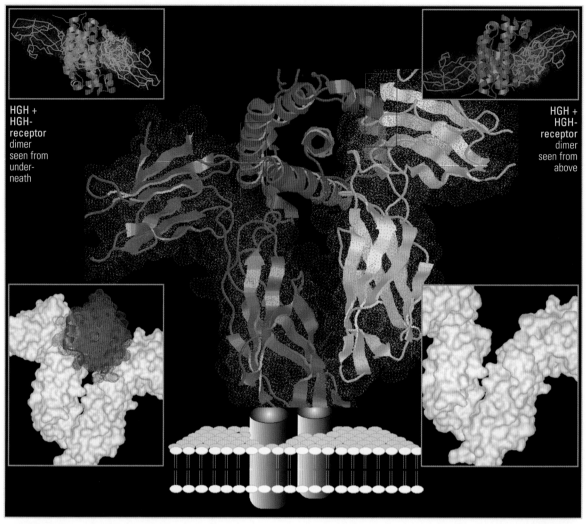

HGH + HGH-receptor dimer seen from underneath

HGH + HGH-receptor dimer seen from above

HGH receptor dimer (in different shades of grey) + HGH monomer, in different views (PDB 3HHR)

Another large group of membrane receptors is the class of single-TMH receptors, which are typical of receptors for *cytokines, growth factors, differentiation factors* and *apoptosis-inducing factors (death factors)* as well as cell *adhesion proteins*. Moreover, also the *B-cell antigen receptor*, the *T-cell receptor* and the proteins of the *major histocompatibility complex* (MHC) are inserted into the membrane with a single TMH (as shown later).

Many of these receptors carry *cytoplasmatic effector domains* with *tyrosine kinase activity* and thus actually represent enzymes although they are not *per se* (= constitutively and permanently) active. Instead they can be activated only through hormone or ligand binding and are therefore called *agonist-dependent* or *-regulated* enzymes, in contrast to other enzymes that are always (= *constitutively*) active. Often there is a need for dimerization, sometimes even trimerization, of the receptors. Growth factors of the *helix bundle cytokine superfamily* each have as a monomer two different receptor contact sites that are sterically appropriately positioned so that one molecule of ligand can asymmetrically dimerize two receptors, as shown here by the example of the human growth hormone (HGH) receptor.

The picture at the bottom right shows interleukin-2 (IL-2) plus its dimerized receptor (PDB 1ILM). The similarity of this complex to the HGH receptor is astonishing. This picture teaches us that signal substances or receptors are not exclusive for a specific area of biology or medicine. Names given to signals generally reflect more the history of discovery than the fundamentals in structure. The *interleukin of immunology* is chemically and structurally of the same kind as the *HGH of endocrinology*, and so are the respective receptors.

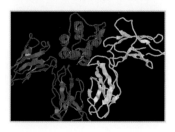

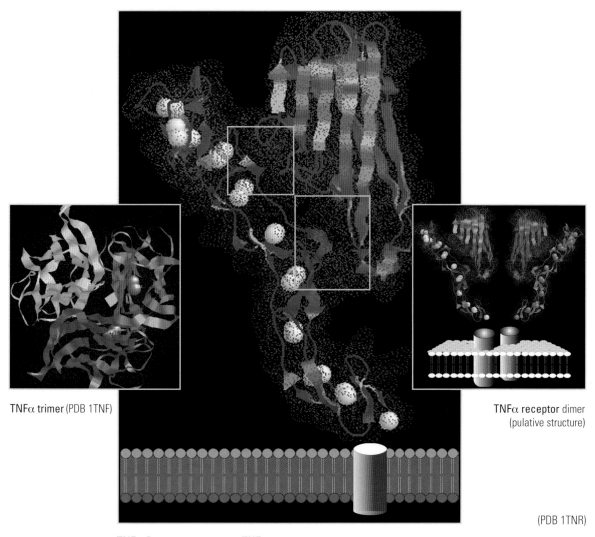

TNFα trimer (PDB 1TNF)

TNFα receptor dimer (pulative structure)

TNFα Receptor monomer + TNFα monomer

(PDB 1TNR)

Other members of the cytokine family (such as interferon) as well as of other growth factor families (such as PDGF) possess only one such contact site and must, therefore, themselves first form a hormone dimer that can symmetrically dimerize two receptors.

Other hormones/signals, such as the death factors TNFα or Fas-ligand, have to trimerize in order to trimerize their corresponding receptors, as shown here with TNFα.

There is a kind of dwarfism called the Laron dwarf in which HGH is normal, but ineffective, since the corresponding receptor in the liver and cartilage/bone-forming cells, due to a mutation in the ECD, cannot recognize HGH as an agonist rendering these cells incapable of responding, i.e. secreting IGF-1. Consequently, augmenting or administering HGH to these patients would be ineffective.

Elimination by apoptosis of lymphocytes that are fortuitously directed against the body's own antigens (proteins, DNA) is a necessary physiological precondition for avoiding autoimmune diseases. Mutations in the Fas-ligand (FasL) or Fas-receptor (FasR), respectively, allow such damaging autoreactive cells to escape timely cell death and to live longer or to become even immortal and thus harmful for the body: they become responsible for diseases such as *systemic lupus erythematosus* or *general lymphoproliferative disease*, respectively.

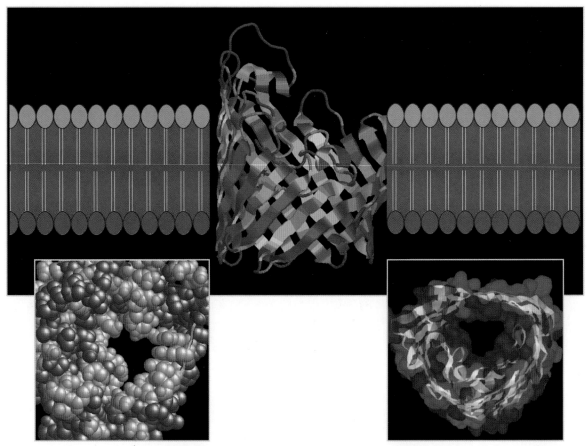

Porin side view, lipophilic amino acids (gray); hydrophilic (light magenta); inset as seen from above   (PDB 3POR)

This picture should teach that there are also membrane proteins which are anchored in the plasma membrane not by α-helices, but by special β-sheet constructions. The inset further shows (viewed from above, i.e. perpendicular to the plasma membrane) that this particular protein has a central hole going through it, i.e. a *channel*; here shown in open conformation. It is also remarkable that the channel is coated exclusively with hydrophilic AAs, and that lipophilic AAs are on the periphery of the protein, the latter thus facilitating its immersion in the lipid bilayer membrane. It is easy to see from this architecture and distribution of AAs that this protein represents a channel for hydrophilic solutes, i.e., water-soluble molecules. It is called a *porin*. Porins are present in the outer membrane of mitochondria (but also bacteria). Porins are able to change their conformation, and hence solute-permeability, in a voltage-dependent manner. There are also ion channels (e.g. specific K^+-channels) anchored in the plasma membrane via β-sheets.

Aquaporins, by contrast, are constructed according to the typical 4 x 6 super bundle principle, i.e. 4 bundles of 6 α-helices each, that applies also to most ion channels.

Aquaporins, however, are generally in the permanently open conformation, as opposed to ion channels that alternate between closed and open states, in either a *voltage-* or *agonist-dependent* fashion. Aquaporins are especially numerous in renal epithelial cells and serve to reabsorb water. Water fluxes are thus regulated only by the number of aquaporins in a given cell. Aquaporins are also present in erythrocytes, which explains why erythrocytes are so much more osmosensitive than other cell types. In these cells, aquaporin represents the *antigen* of the COLTON blood group.

> Somewhat surprisingly, mutations in an aquaporin gene can also be responsible for a form of retinitis pigmentosa (type 9), which can lead to blindness.

The crystal structure of most membrane proteins has not yet been elucidated simply because these proteins, due to their lipophilic membrane domains, cannot be crystallized. However, the hydrophilic extra- as well as intracellular domains can be split, or even better, can be individually expressed as recombinant proteins and, as such, be crystallized.

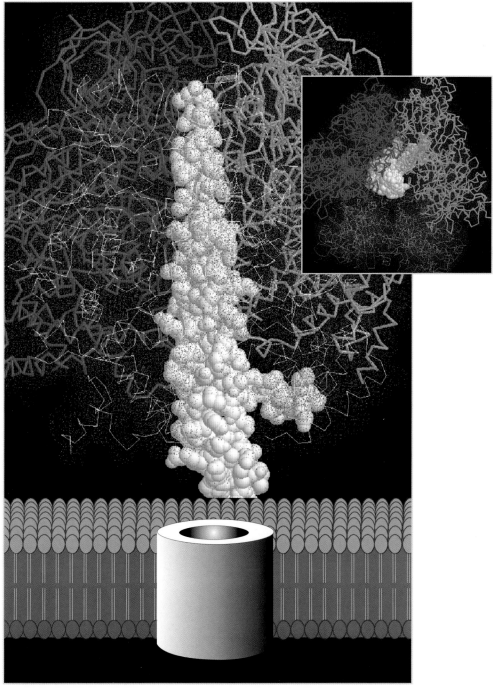

F1-ATPase from mitochondria (PDB 1BMF)

At the end of this chapter, a highly complex membrane protein should be shown, i.e. the proton-pumping ATP synthase. The picture above shows the inner mitochondrial membrane seen from the intermembrane space (i.e. the fluid compartment between the outer and inner mitochodrial membranes). The transmembrane portions (not modeled) represents the so-called F0 unit, a H^+-pumping channel, that is connected with the F1 unit, which – as shown – is a heterohexameric protein consisting of 3 α and 3 β subunits (shown backbone and dots representation and in different colors). In the center of F1, there lies the γ subunit, which is a coiled coil helix that connects F1 to F0, the so-called stalk. The insert to the right shows an axial view of F1 from above. This ATP synthase couples proton transport with the synthesis of ATP.

Signal Transduction Proteins

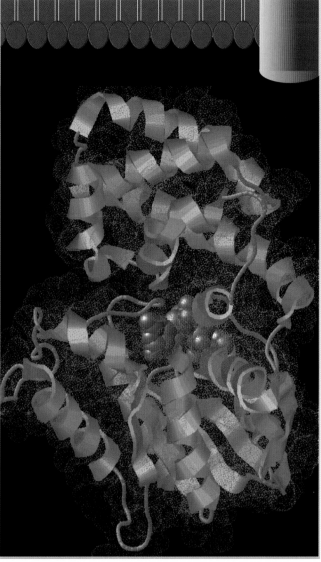

Giα + GTPγS (inhibitory GTP-binding protein α-subunit)   (PDB 1GIA)

As explained in the previous chapter, hydrophilic signals can exert their effects on a cell only via membrane receptors, i.e. by inducing a conformational change in the receptor upon binding, and so make, in a certain sense, their way to the cytoplasmatic side, since the receptor itself is membrane-anchored. The seven transmembrane helix receptors (7TMHR) then become competent to interact with their cytoplasmatic loops and C-terminal ends (as a whole called the *coupling domain*) with G-proteins. G-proteins are α:β:γ-trimers, which are partly anchored by a fatty acid residue on the inside of the membrane, and thereby kept in close proximity to the 7TMHR. A signal-activated 7TMHR couples with an α-subunit, which then undergoes a conformational change, thereby releasing a previously bound GDP and permitting GTP to enter the same pocket. This leads to a further conformational change that results in the dissociation of the GTP:α-subunit from the receptor on one side and of the β:γ subunit on the other. The released GTP:α-subunit is now in a position to undertake a discrete change of place and look for a target protein. Such proteins are abundantly present in every cell type, i.e., in adequate density, and thus always in close proximity to the receptor, e.g. an effector enzyme such as *adenylate cyclase* or *phospholipase C* or *phospholipase A2* or an *ion channel*, respectively.

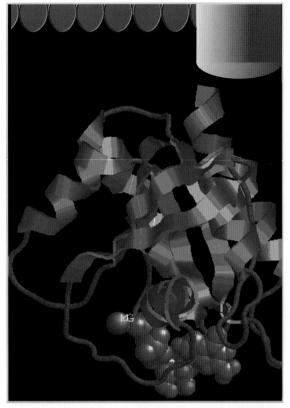
Ras p21 protein + GDP (PDB 1CRQ)

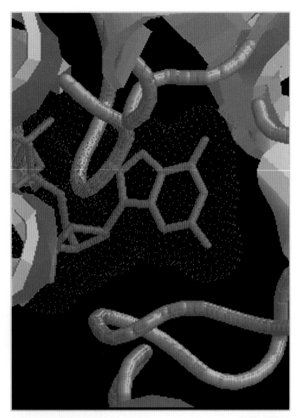
Transducin α-subunit + GDP (PDB 1TAD)

Each of these target proteins is again an integral membrane protein with intracellular contact sites with which an active GTP:α-subunit can interact. Of course, an α-subunit also possesses a contact site specific for one given target protein. Each of the latter is basically in an *inactive* or *closed* conformational state, respectively. It is the interaction with the GTP:α-subunit that induces, again, an appropriate conformational change in that target protein, which activates it, as in the case of an enzyme, or leads to an open state, as in the case of an ion channel.

The activated adenylate cyclase can thereupon convert ATP into cAMP and release it into the cytosol. This is effectuated with a certain speed so that during the active period, many hundreds of cAMPs are produced. Thus, the primary signal not only becomes transmitted, but amplified. Cyclic AMP is thus called a *second messenger* that, soluble in the cytosol, can move freely and, as a small ligand, can collide with target proteins further downstream.

One of these is the cAMP-dependent protein kinase (PKA) that, when so activated, can phosphorylate suitable target proteins on serine or threonine residues. Thus phosphorylated, these proteins become competent to execute their intrinsic activities, e.g. to become involved as an active kinase in intermediary metabolism, or to promote steroid synthesis as an activated key enzyme, or to regulate gene expression as an activated transcription factor, and so on. These phenomena represent the transition from *signal transduction* to the actual *effector apparatus* of a cell.

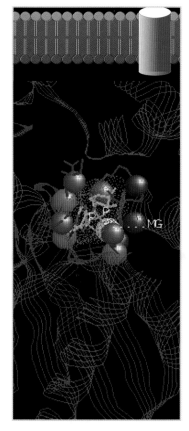
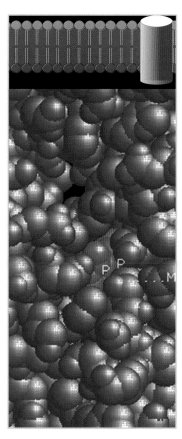

Gsα ± GDP + Mg^{2+} (stimulatory GTP-binding protein α-subunit) (PDB 1GFA)

GTP:α-subunit-activated phospholipase C or phospholipase A2 catalyze the release of diacylglycerol (DG) and inositol trisphosphate (IP3) or arachidonic acid from membrane phospholipids (phosphoinositol bisphosphate, PIP2), respectively. DG and IP3 function as a further set of second messengers and activate the DG-dependent protein kinase C (PKC) or open the IP3-dependent Ca^{2+} channel of the endoplasmic reticulum, the latter resulting in a transient increase in free cytosolic Ca^{2+}, which again activates, as second messengers, a series of Ca^{2+}-dependent proteins, e.g. calmodulin or,

in the case of excessive increase in Ca^{2+}, the cell-damaging proteases (neurotoxicity) mentioned previously.

An ion channel for K$^+$ ions, opened by a GTP:α-subunit, would, depending on the extracellular gradient, permit intracellular K$^+$ ions to flow out, resulting in hyperpolarization of the cell and thus reduced sensitivity to depolarizing stimuli. This would be the basis for the previously mentioned pain inhibition through opiates. The release of pain transmitting, activating depolarizing neurotransmitters in the afferent sensory neurons is presynaptically hindered. Arachidonic acid, in contrast, is not a second messenger, but rather a substrate for lipoxygenase as well as cyclooxygenase. These enzymes thus generate secretory hormones such as leukotrienes or prostaglandins, prostacyclins and thromboxanes, respectively, all of which exert their effects via 7TMHRs on neighboring cells.

Prostaglandins play a role in inflammatory processes and therefore pain resulting from inflammation. Headache can generally be successfully treated by drugs that inhibit cyclooxygenase. Acetylsalicylic acid (aspirin) represents an appropriate chemical for this purpose. As a small ligand, it fits into a pocket of the enzyme and can transfer its own acetyl residue to serine 530 within the enzyme's active center. Thus, through covalent modification of a residue essential for catalysis, the enzyme becomes inactivated for a long time.

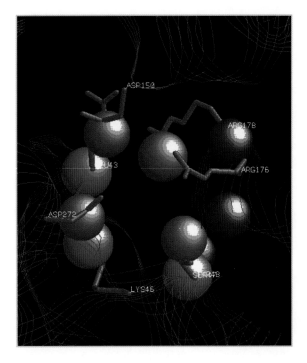
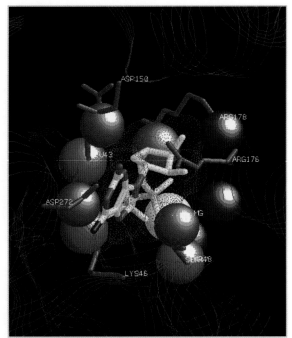
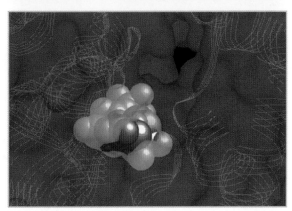

Gsα ± GDP + Mg^{2+} (stimulatory GTP-binding protein α-subunit) Left panel: the GTP-binding pocket empty, unliganded, the GTP-contacting atoms of the protein highlighted by CPKs; right panel: the GTP-binding pocket filled, liganded with GDP.

(PDB 1GFA)

There are several different G-proteins. They differ, however, only in their α-subunits, not in their β:γ-subunits (strictly speaking, there are also a few iso-β and iso-γ subunits). Analogously, but conversely to the glycoprotein hormones (four different β-subunits together with one single α-subunit resulting in four different α:β-holohormones), different G-protein α-subunits combined with the same β- and γ-subunits result in different α:β:γ trimers that are membrane-associated and 7TMHR-contacting. Different α-subunits have, on the one hand, a preference for different 7TMHRs, and on the other hand, specificity for different effector/target proteins. Furthermore, different α-subunits have differing susceptibilities to opportunistic ligands or poisons, such as cholera (CTX) or pertussis toxin (PTX).

For example, there is a **Gsα** that contacts the β-adrenergic 7TMHR on one side and, in **s**timulatory fashion, adenylate cyclase on the other, or there is a **Giα**, which contacts an α$_2$-, but not an α$_1$-adrenergic 7TMHR on one side, and adenylate cyclase on the other, but in an **i**nhibitory fashion. There is also a **Gpα** that contacts an α1, but not a β-, and an α$_2$-adrenergic 7TMHR on one side, and **p**hospholipase C on the other, or there is a **Gkα** that contacts the opiate 7TMHR on one side and **K**$^+$ channels or Ca^{2+} channels on the other, opening the former and closing the latter, respectively. There is also **Goα** that contacts the **o**lfactory 7TMHR on one side and, in a stimulatory fashion, the adenylate cyclase on the other. In retinal cells, there is **Gtα** (**t**ransducin) that contacts the visual 7TMHR (rhodopsin and opsins) on one side and, in a stimulatory way, the cGMP phosphodiesterase. There are a few more Gαs.

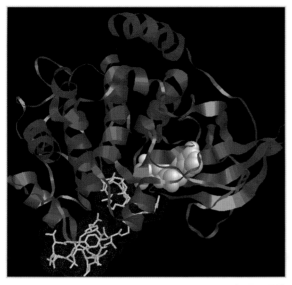

Protein kinase A + ATP (white) + inhibitory pseudosubstrate peptide (magenta) (PDB 1 ATP)

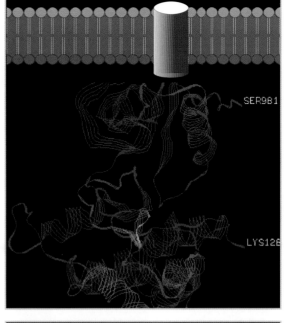

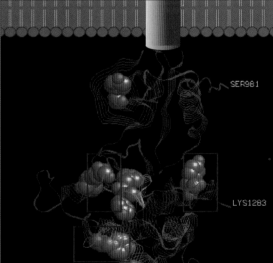

Insulin receptor cytoplasmatic tyrosine kinase domain, i.e. the AAs 981-1283 of the entire receptor protein. These three pictures show (above right) the backbone of the protein as the C-terminal part below the single TMH of the insulin receptor. Below left, the tyrosine residues are highlighted as CPKs. It can clearly be seen that the majority of the tyrosines are found in the lower subdomain reaching deeper into the cytosol, and that it is these that will potentially be autophosphorylated. The picture below on the right side shows, however, that not all of these tyrosines are capable of being phosphorylated. Some are pointing inwards and are thus not accessible. In the lower left picture, inaccessible tyrosines, i.e. masked by other AAs, are indicated with a frame.

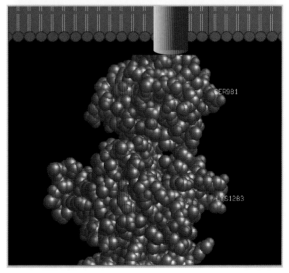

Insulin receptor tyrosine kinase domain (PDB 1IRK)

Signal Transduction Proteins

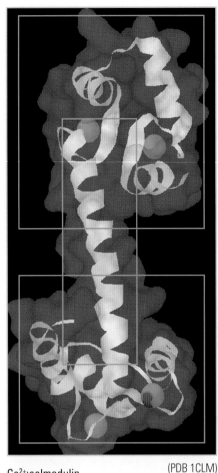

Ca^{2+}:calmodulin (PDB 1CLM)
(including four Ca^{2+} ions)

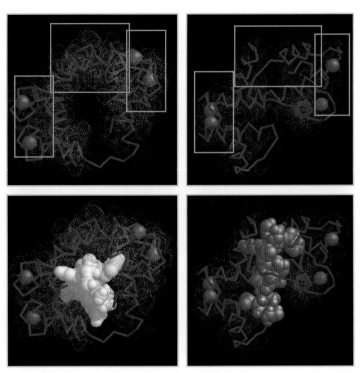

Ca^{2+}:calmodulin (including four Ca^{2+} ions) in its conformation (PDB 1CDL)
competent to bind the calmodulin-binding helix of smooth
muscle myosin light chain kinase (MLCK), in axial view (left panels)
and in longitudinal view (right panels) without MLCK (upper panels)
and with MLCK (lower panels)

In addition to cAMP, DG and IP3, ionized calcium (Ca^{2+}) is, as previously mentioned, an important second messenger. It is among the smallest of biomolecules, although its principle of action as a ligand is the same as that of the other larger molecules, namely binding to a target protein and thus allosterically activating it. One of the many Ca^{2+}-dependent proteins is calmodulin which possesses a total of four Ca^{2+} pockets that can be sequentially liganded. With increasing liganding, calmodulin undergoes several discrete *micro*conformational changes that finally lead, in the presence of an appropriate, i.e. a calmodulin-dependent, target protein, to a strong bending in the helical midregion of calmodulin (white frame), a *macro*conformational change that enables calmodulin to intimately contact the target protein and thereby to activate it.

Above right, an example of myosin light chain kinase (MLCK) of the smooth muscle shows how the MLCK's calmodulin-binding domain is engulfed by bent 4Ca^{2+} calmodulin. The activated enzyme phosphorylates the myosin light chain that, in the phosphorylated form, undergoes a conformational change in the sense of a shortening of its overall length.

Thereby, the entire myocyte also contracts. It can be clearly seen that the bent calmodulin forms a large central pocket by having brought its two globular 'heads' (framed in magenta) next to each other. The same protein can thus possess many small as well as large pockets and thus bind many ligands of different sizes such as 4 x Ca^{2+} and 1 x MLCK.

There are several Ca^{2+} as well as Ca^{2+}:calmodulin (CAM)-dependent proteins. The former class includes *troponin C* (in skeletal muscle cells), which is very similar to calmodulin. Other Ca^{2+}-dependent proteins are *α-actinin, calcineurin B, calretinin, caltractin, β- and γ-crystalline* of the eye, IP3- and DG-generating *phospholipase C, protein kinase Cα, myeloperoxidase, parvalbumin, S-100 protein* (involved in cell cycle progression and cytoskeletal-membrane interactions) among others. All these proteins possess structurally very similar Ca^{2+} pockets (as shown at the beginning of this book), but there are also proteins that are sensitive to extracellular Ca^{2+}, such as the *Ca^{2+}-sensing receptor* of the parathyroid gland which possesses a distinctively different Ca^{2+} pocket in its extracellular domain.

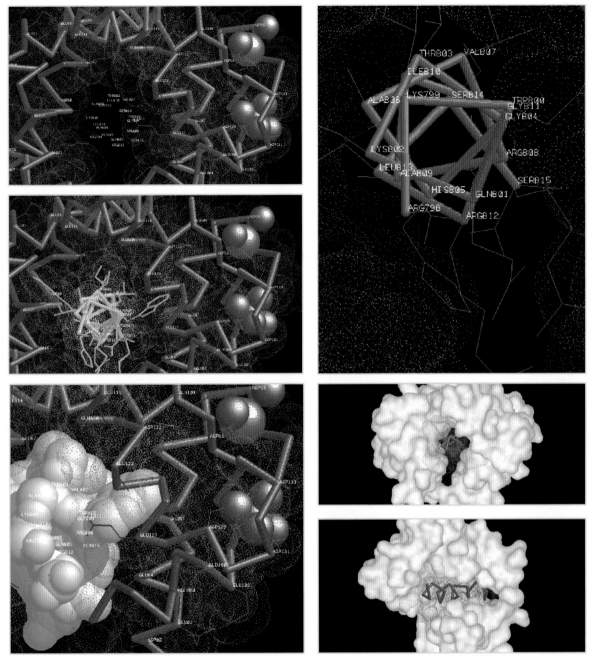

Ca²⁺:calmodulin (including four Ca²⁺ ions) complexed with the calmodulin-binding helix of **MLCK** (PDB 1CDL)

CAM-kinase (a protein kinase), cAMP/cGMP-phosphodiesterase, and Ca^{2+}-ATPases are examples of the class of Ca^{2+}:calmodulin (CAM)-dependent proteins, all of which possess a common large calmodulin interaction domain (and not a small Ca^{2+} pocket). This is a helix, shown here enlarged with MLCK as an example, with amphiphilic character. The upper portion of the α-helix, shown here axially, contains predominantly lipophilic amino acids (AAs), while the middle portion contains primarily basic hydrophilic AAs (Lys, Arg, His). Correspondingly, the lower portion of the counterpart, i.e. the large *groove* in calmodulin, is coated with hydrophilic and negatively charged AA residues, which thus enable a very high-affinity electrostatic attraction of the target protein's α-helix to Ca^{2+}:calmodulin. The groove of calmodulin possesses not only a 'macroscopic form', i.e. an ellipsoid channel, but additionally microscopic wall characteristics. All these properties have to be considered under the concept of *'complementarity'* (compare the explanation of *'patches'* in the chapter on 'TGFβ' earlier in this book).

Signal Transduction Proteins

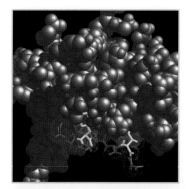 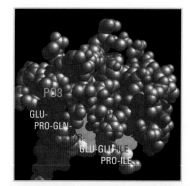 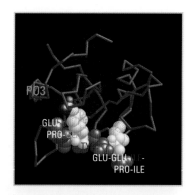

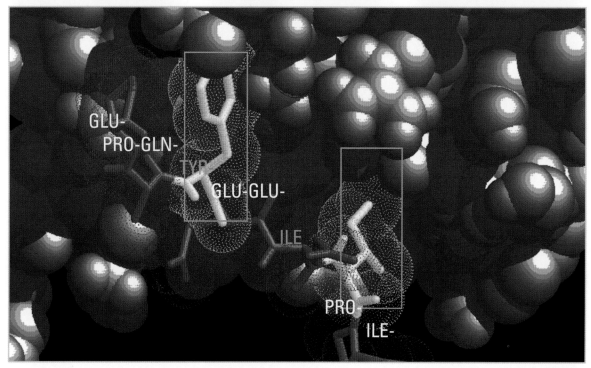

Src SH2 domain + phosphopeptide Glu-Pro-Gln-Tyr-Glu-Glu-Ile-Pro-Ile (PDB 1SPS)

Many intracellular signal transduction pathways utilize protein kinases. Some of these have tyrosine substrate specificity. Phosphotyrosines themselves, when situated within a context of Pro-Gln-**Tyr**-Glu-Glu-**Ile** serve as docking sites for further adapter tyrosine kinases and other signal transduction proteins, provided the latter possess a corresponding *cleft*, i.e. a so-called Rous sarcoma virus **S**RC **h**omology (**SH**)2 domain.

Instead of explaining the pictures above by myself, I let the authors from the Rockefeller University in New York speak and the reader decide, how these images match their words:

*'The crystal structure of the Src SH2 domain complexed with a high affinity 11-residue (Glu-**Pro-Gln**- **Tyr** Glu-Glu-**Ile**-Pro-Ile-Tyr-Leu) phosphopeptide has been determined at 2.7 Å resolution by X-ray diffraction. The peptide binds in an extended conformation and makes primary interactions with the SH2 domain at six central residues:* Pro-Gln-**Tyr**-Glu-Glu-**Ile**. *The phosphotyrosine and the isoleucine are tightly bound by two well-defined pockets on the protein surface, resulting in a complex that resembles a two-pronged plug engaging a two-holed socket. The glutamate residues are in solvent-exposed environments in the vicinity of basic side chains of the SH2 domain, and the two N-terminal residues cap the phosphotyrosine-binding site.'*

[Waksman G, Shoelson SE, Pant N, et al: Binding of a high affinity phosphotyrosyl peptide to the Src SH2 domain: Crystal structures of the complexed and peptide-free forms. *Cell* 1993;72:779–790.]

Multimodular Adhesion Proteins

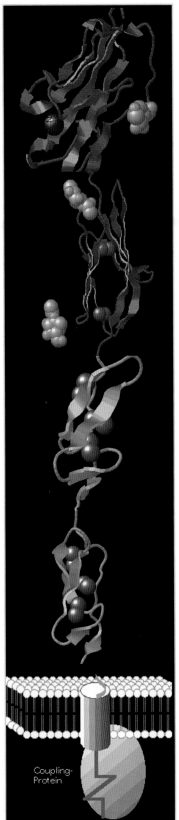

Ig-like module

Immunoglobulin (Ig)-like module (glycosylated: CPKs) (PDB 1CDB)

EGF-like module

Epidermal growth factor (EGF)-like module (γ-sulfur atoms of disulfide bridges in CPKs) (PDB 1APO)

Transmembrane helix and coupling domain with signal transduction protein

Coupling-Protein

Cells must be able to recognize not only soluble signals *swimming* freely in the extracellular fluid, but also insoluble signals, i.e. those anchored in the membrane of neighboring cells. Such signals are called adhesion ligands and the corresponding recognition proteins adhesion receptors. Often, both are similar in size or are even one and the same protein (homotypical interaction) so that one cannot tell which one is the ligand and which the receptor. That is why the cell-cell interaction molecules are simply called adhesion *proteins*. A homotypical interaction follows the general principle of protein dimerization. Adhesion proteins can, in addition, contact proteins of the extracellular matrix. Adhesion proteins are primarily anchored in the membrane with a transmembrane α-helix and, like other receptors, possess a cytoplasmatic coupling domain that allows interaction with intracellular signal transduction proteins or cytoskeletal proteins, mostly in a ligand-dependent manner.

Since the next cell is, in molecular terms, relatively far away, the extracellular domains of adhesion proteins have to be quite long to bridge that gap. Evolution solved this problem by linking together several modules in a linear rather than a convoluted (globular) fashion (compare what was said earlier about the ECD of the hCG receptor). These modules are themselves relatively small proteins (50–200 amino acids), stable and folded in a compact manner (*independent folding domain*). Identical as well as different modules can be concatenated to form a variety of proteins of different lengths, interaction specificities and other functions.

On the right is such an adhesion protein represented as a model. Its ECD consists of two EGF-modules (turned in different angles in the y-axis) and two immunoglobulin (Ig)-like modules, also turned 90° relative to each other. The latter are, in addition, glycosylated (shown as CPKs).

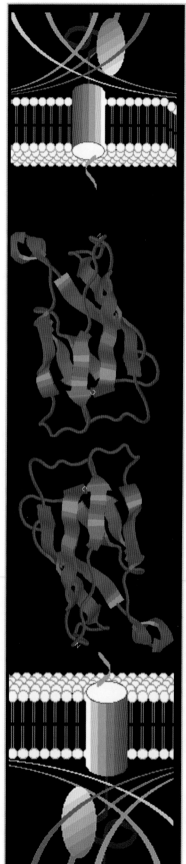

Coupling domain with cytoskeleton and adaptor proteins

Cadherin module (PDB 1NCG)

Cadherin module (PDB 1NCG)

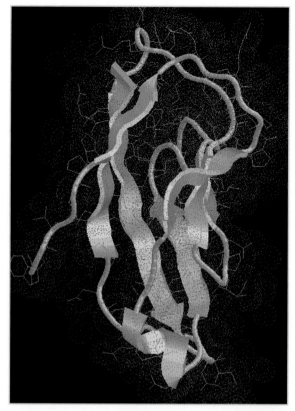

Cadherin module (PDB 1NCG)

Here again, a cell-cell contact is shown as a model, based on a homotypical interaction of two identical adhesion proteins. Each possesses a long extracellular domain comprising several cadherin modules (for the sake of simplicity, only one of each on each side is depicted).

> Cadherin contacts are essential for keeping identical cells together in an organ. If a tumor develops in an organ, the tumor cells also initially remain in close contact and thus resident within the organ. This is the case for a local, organ-confined and thus generally well resectable (*benign*) tumor. Through a mutation in a gene for such a cadherin adhesion protein, the tumor cells, however, can lose their ability to contact each other. As a consequence, they can escape from where they were originally embedded and spread out into the surrounding tissue, even travelling to distant parts of the body where they cause metastases (daughter tumors). By this property, the tumor is characterized as being *malignant*.

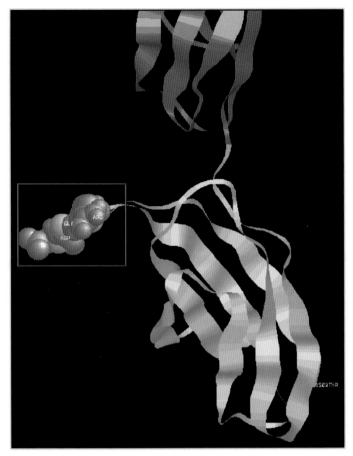

Fibronection type-1 cell adhesion module with **RGD** (in CPKs) motif

(PDB 1FBR)

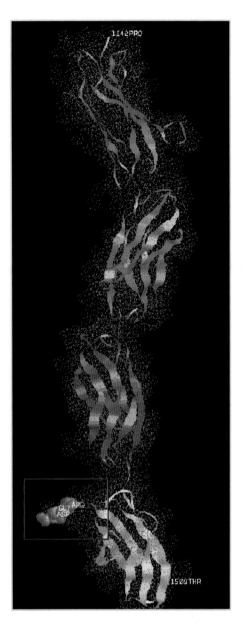

Cell adhesion proteins can interact, as previously mentioned, not only with cells, but also with proteins of the extracellular matrix (ECM). These, too, consist of repetitive modules. Proteins of the ECM belong to the class of the largest proteins (> 2,000 amino acids long). Like a fibrous network, different ECM proteins criss-cross over each other, bind to each other at certain contact points, most densely in the basal membrane of epithelial tissues, and thus build a network or matrix to which cells can adhere. There are several isofibronectins and different fibronectin modules.

On the right is a picture of such a filament that consists of four fibronectin type-1 modules. An interesting structure jutting out of this filament is the RGD motif that, in addition, is physicochemically quite distinct. It consists of the AAs arginine (R), glycine (G) and aspartic acid (D), which provide in a small space a characteristic and dense mixture of positive (R) and negative (D) charges, thus forming a highly reactive mini-domain. Actually, it is the RGD-motif that is the *ligand* within the whole protein that *docks* to the adhesion receptor (an *integrin*). RGD sequences are also found in many other ECM-proteins (vitronectin, laminin, collagen, von Willebrand factor, etc.).

Mutations as well as autoantibodies against ECM proteins can lead to severe diseases, such as psoriasis (skin blisters), Goodpasture's syndrome (kidney failure) and others.

DNA-Binding Proteins

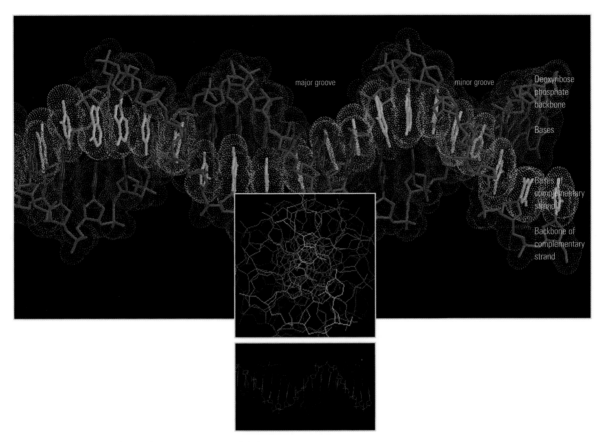

Double stranded DNA (dsDNA), longitudinal and axial views

Deoxyribonucleic acid (DNA) is the chemical basis of genetic information, i.e. the *gene*. Human beings possess about 30,000 genes (the *genome*). Genes code for proteins and it is the proteins that make life possible. Enzymes, signals, signal receptors, signal transduction proteins, and cell matrix proteins are examples of vital proteins that have been discussed thus far. Remarkably, a cell can be viable even when it no longer possesses any DNA (e.g. red blood cells) or when DNA is compacted into chromosomes (visible only under a microscope) so that any transcriptional activity is rendered impossible. But a cell can never be healthy or viable even if only one of the roughly 20,000 different proteins it expresses (the *proteome*) is missing or defective.

DNA is a double strand in which bases are sequentially linked to a deoxyribose-phosphate backbone, comparable to the AAs in proteins, with one important difference: The bases here always points inward, unlike the side chains of amino acids in an α-helix of a protein, which point outward (as shown above). Bases that face each other tend to hybridize and thus hold together the two DNA strands via hydrogen bonding, forming a winding double helix. There are only four different bases: adenine (A), thymine (T), cytosine (C), and guanine (G).

Only specific bases can together form a basepair: A + T and C + G. The synthesis of DNA requires specific proteins, i.e. a *DNA polymerase*, among others. Prior to every cell division, the entire DNA must be duplicated. In programmed cell death (*apoptosis*), enzymes are activated that cut the long DNA into pieces. Of course, such *endo-* and *exonucleases* must not be constitutively active, but must become active only in a strictly regulated fashion, i.e. by *death factors* that induce apoptosis. Bacteria possess endonucleases of very different cutting specificities in order to destroy the DNA of bacteriophages, i.e., viruses that have intruded into them. *DNA-dependent RNA polymerase* transcribes a piece of DNA, i.e. the structural or coding region, of a gene coding for a specific protein, similar to how DNA-polymerase synthesizes a new (opposite) strand from the *matrix* or *template* strand of DNA.

In all species, the genetic information is stored in the form of DNA. This is the case in bacteria as well as most viruses, with the exception of retroviruses, in which RNA represents the structure of genes. Retroviruses therefore must bring into the affected host cells an enzyme that transcribes their RNA into DNA, the so-called *reverse transcriptase*.

> One of the most important retroviruses is HIV, which causes AIDS.

Genes (materially: DNA) code for *proteins* in that a specific sequence of three bases (a base triplet) stands for a specific AA (amino acid). Bases can be altered by various agents (UV radiation, chemical mutagens).

> The consequence is a mutated gene leading to the expression of a *mutated protein* and thus, potentially, to *disease*. Many such DNA damages can be *repaired*, but some can not. When a *DNA repair enzyme* is itself damaged by mutation, the DNA damage remains unrepaired and can lead to early cell death, to a disease called progeria (premature aging in human beings), or to tumor development.

A gene contains not only *coding sequences* (in the *exons*) but also noncoding sequences. The latter, called *introns*, are situated in an intervening manner between the exons. In addition, there are many more non-coding sequences in the *upstream regulatory region* of a gene, that often by far exceed in length all exons together. These sequences are required to decide *when*, *in which cell*, *how much*, and under *which* external or cell-internal *influences*, the gene concerned is expressed.

> Mutations in this region do not lead to a defective protein, but to a protein being overexpressed or underexpressed, or expressed ectopically (i.e. in cells where it is normally not expressed). Dys- or deregulated genes are also responsible for diseases, very often cancer.

Proteins that regulate expression are called *transcription factors*. They interact with specific sites (*cis elements, response elements*), consisting of a few basepairs within the regulatory region of a gene, and thus decide whether and how often the coding gene region located downstream of this regulatory region (the *trans*-gene) is transcribed into RNA. There are at least two possible ways in which molecules can interact with DNA.

(1) *Small molecules* can fit perfectly between the bases of the two strands provided they are sufficiently flat, as explained earlier using *anthracen* and *mitomycin* as representatives of other *mutagens* and *cytostatic* agents.

(2) *Large molecules*, i.e. proteins, in contrast, surround both DNA strands from the outside for longer or shorter stretches, depending on the size of the protein itself, and so interact with a larger or smaller number of bases from either strand.

Furthermore, DNA-binding proteins are generally quite specific, i.e., they do not bind indiscriminately to DNA at any point, but literally search for (by *sliding* and *hopping*) and, when found, bind to *specific base sequences* within the entire DNA, only to these and to no other. In contrast, small molecules primarily contact one or two bases and can, therefore, be deposited at thousands/millions of different places within the DNA. Recognition sequences for proteins range from 5 to 50 bases, and sequences of such length are represented in the entire DNA of a cell very infrequently or may be even unique.

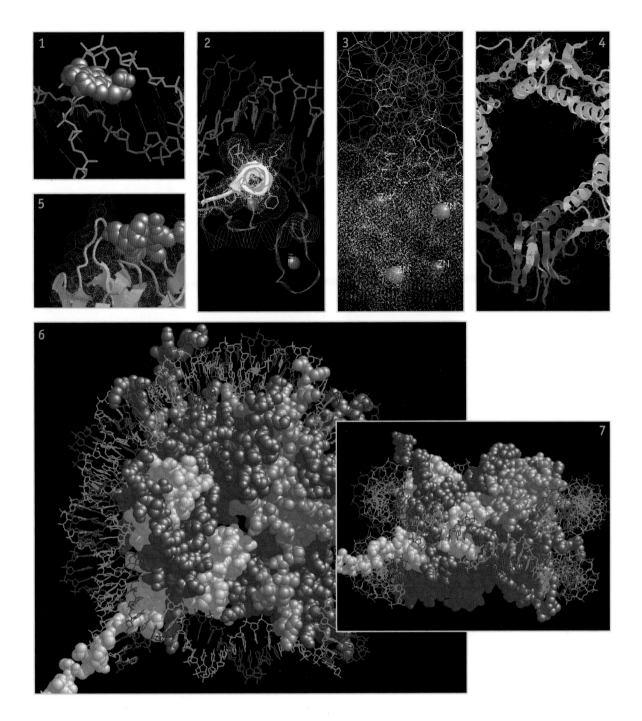

Different Patterns of Protein – DNA Interaction

(1) Small molecular toxins, e.g. mitomycin (PDB 199D), bind to dsDNA from inside by intercalating between the two strands of dsDNA (panel 1).

(2) Transcription factors and other DNA-binding proteins, e.g. the glucocorticoid receptor (PDB 1GLU), or the sliding clamp of DNA polymerase (PDB 1B77), bind from outside to dsDNA by partially or totally engulfing it (panels 2, 3, 4).

(3) Similarly, but inversely, dsDNA can also bind as a superhelical coil around a large protein complex, e.g. the nucleosome complex (PDB 1AOI) consisting of H2A.H2B.H3.H4 histone proteins (panels 6, 7).

(4) Oligonucleotides and other small DNA fragments can be engulfed more or less totally by an antibody, such as seen with TTT in panel 5 (PDB 1CBV).

DNA/RNA-Processing Enzymes

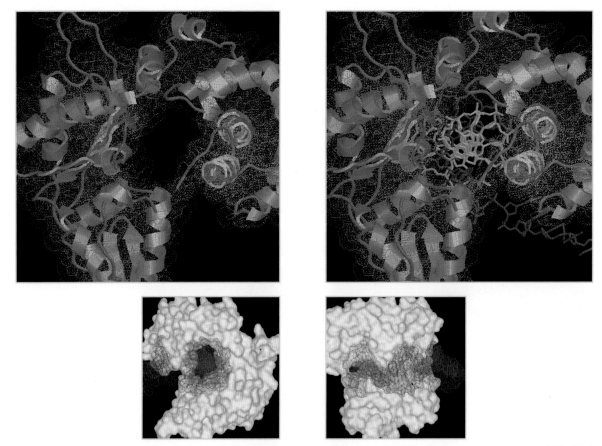

Human DNA polymerase empty (left upper panel) and filled with gapped DNA (right upper panel) (PDB 1BPX)

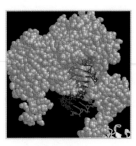 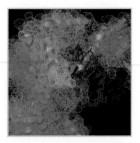 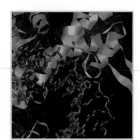 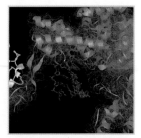

T4-bacteriophage DNA polymerase + DNA in different views (PDB 1T7P)

It can clearly be seen from the pictures above and those that follow that enzymes are much larger than the axial diameter of dsDNA. Therefore, such enzymes can engulf a certain stretch of dsDNA longitudinally. The presence or development within the protein of a groove of appropriate size and complementarity, tailored to the respective surface of that stretch of dsDNA, is prerequisite for binding to substrate DNA. The binding, on the other hand, must not be static; instead, there is need for a further spatial processing of the enzyme relative to the substrate, or, looked at from the opposite side, the pushing out of the product from the enzyme. For this, cooperation with energy-consuming 'motor' proteins is required.

Mutations in the gene of a *DNA polymerase* are, of course, incompatible with life. Therefore, there has never been a patient seen with this mutation. There are, however, patients with mutations in *DNA repair enzymes*, resulting in *progeria* (premature aging and death) among several others diseases, as mentioned earlier.

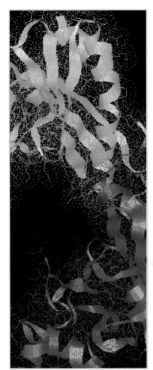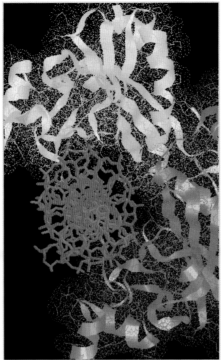

Bacterial *Eco*Rv endonuclease dimer ± dsDNA (frame in right picture indicating the cutting point) (PDB 2RVE)

The picture above and those that follow show that a suitable size for establishment of a DNA-binding *groove* of a protein generally requires *homodimerization* of the respective protein, mostly asymmetrically (monomers shown here in different colors).

Bacterial endonucleases represent the most important reagents of today's *molecular biology*. With their help, particularly with the cleavage site-specificity inherent to such enzymes, it became possible to produce *dsDNA fragments* of predictable lengths and thus diagnose *mutations* in patients' DNA, or to use these fragments to ligate DNA segments of a gene into a *vector* that then can be used to *transfect* cells and produce *recombinant proteins*, or generate *transgenic animals* (generally mice) with specific *genetic defects* to serve as models to study the function of the gene (protein) in the corresponding human disease. All this opened the way for gene therapy of humans with a disease which came into the realm of today's medicine.

In 2001, the entire human genome has been sequenced (*Human Genome Project*) and approximately 30.000 genes described. It has already become possible to repair some defects in human somatic cells (*gene therapy*)! None of the goals, methods, and strategies mentioned above would be possible without the use of bacterial endonucleases.

HIV reverse transcriptase has already been presented in the Chapter 'Enzymes' (on page 38).

DNA/RNA-Processing Enzymes

Antibodies against DNA

A short piece of ss-DNA is small enough to fit entirely into the antigen-binding region (variable region) of an antibody. An antibody, in contrast to an enzyme or a transcription factor (see later), cannot enclose a DNA double helix, but can, because of the relative smallness of its Fab portion, only bind to a small, i.e. a few nucleotides-long segment of the DNA, in a superficial rather than engulfing way. In the picture above left, one can see the respective N-terminal Ig (immunoglobulin module) of the antibody (magenta = light chain; grey = heavy chain). The loops of the hypervariable regions of the antibody, which constitute the *bed* for the antigen or the epitope of a large antigen, can be clearly recognized.

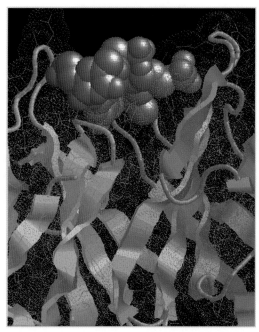

Fab portion of an antibody + trinucleotide TTT (PDB 1CBV)

Anti-DNA antibodies are found in patients with systemic lupus erythematosus and are thus essential elements to diagnose this autoimmune disease.

DNA-Bending Proteins (HMG-Box Proteins)

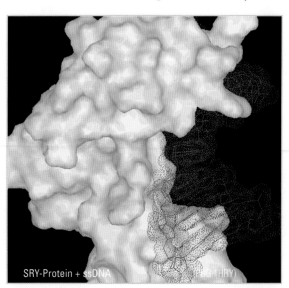

SRY-Protein + ssDNA (PDB 1HRY)

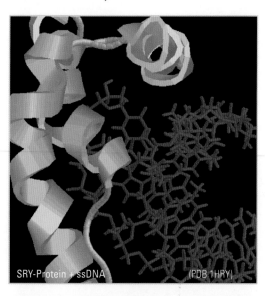

SRY-Protein + ssDNA (PDB 1HRY)

The picture on the right shows how DNA-binding helices (grey, cartoon) of the human SRY protein surround a piece of ssDNA and can cause it to bend. Such *DNA-bending* in the promoter region of a gene appears to be essential for efficient transcription. HMG (high mobility group) box proteins can execute such DNA bending and the SRY protein is one of the HMG-box proteins.

The *SRY protein* is coded by a gene located on the Y chromosome and is prerequisite for male normal sexual development. Mutations in this gene cause male hermaphroditism, i.e. an individual who carries the male chromosome set (44, X, Y) but, both externally and internally, has a female phenotype. Hence, such individuals are considered female from a legal point of view.

Transcription Factors

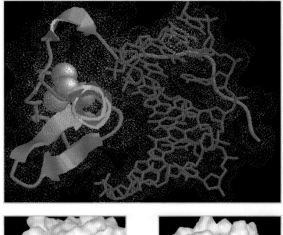
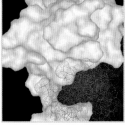
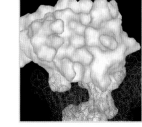

GATA-binding factor (PDB 1GAT)

POU-domain factor **Pit1** (PDB 1AU7)

Transcription factors bind to 5–50 basepair-long domains within the *promoter* region of a gene. Such domains constitute a stretch of quite a unique sequence *(motif)*. They are also called *cis elements for transactivating factors* or, generally, *recognition sequences for transcription factors*. They are designated *response elements* (RE) when the transcription factors binding to them are not constitutively active (i.e. competent to function as transcription regulators) but dependent upon external signals. Such *signals* mediate a *genomic response* in the target cell, via a specific transcription factor, by *up-* or *downregulating* the expression of a specific gene. Signals can interact directly or indirectly with a particular transcription factor in the sense of an allosteric activator.

Steroid and *thyroid* hormones do so directly as they can diffuse through the cell membrane by virtue of their lipophilic character or may, even protein-bound, be carried through by transporters.

Hydrophilic signals, in contrast, would never make their way into the cell through the membrane, but instead react with membrane receptors thus setting in motion the intracellular signal transduction cascades that, for example, activate a protein kinase which, in turn, phosphorylates (among other targets) a transcription factor and thereby activates (or deactivates) it.

A third group of transcription factors is constitutively active, i.e., is always capable of regulating transcription. Among them are the cell-specific master regulatory factors. It is not their activity that is regulated, but rather the time of their appearance, which, in turn, depends on the developmental (maturity) and differentiation stage of a given cell type. These factors are generally effective only permissively, i.e. require for the actual expression of a gene that yet another, mostly signal-dependent transcription factor, binds simultaneously to the same gene promoter on a different site (i.e. response element).

GATA proteins are typical for hematopoetic cell types. A defect in a GATA protein can lead to disturbances in the maturation of such cells, e.g. erythrocytes. The GATA-binding motif itself can also be mutated, e.g. in the promoter of a gene for the *Duffy blood group antigen*. The product of the Duffy gene is, in fact, functionally a *chemokine receptor*, that is occasionally misused by *Plasmodium vivax*, as a docking receptor to gain entry into a cell and cause malaria. Obviously, this parasite carries a surface protein, a domain of which functions as an opportunistic ligand for this chemokine receptor. Individuals with this GATA motif mutation are resistant to malaria because their erythrocytes are negative for the Duffy protein!

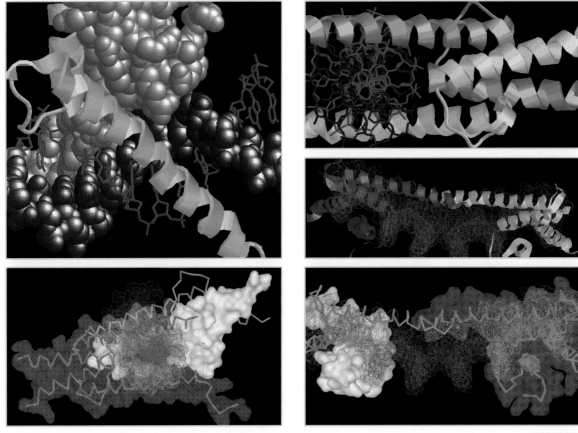

MyoD dimer + dsDNA, helix-loop-helix-motif in MyoD dimer (PDB 1MDY)

The pictures above show *MyoD*, i.e. longitudinally a piece of dsDNA complexed with a MyoD homodimer, of which, again, one can see only the actual DNA interacting segments, i.e., a helix (one monomer grey, the other light magenta). As seen from this longitudinal view of the protein(s), both helices lie in opposite *major grooves* of the dsDNA and so surround dsDNA in a forceps-like fashion. The correct orientation of these DNA-binding helices within a MyoD molecule is brought about by the cross-wise positioning of the two homodimerizing helices, one on top the other (picture right).

MyoD belongs to the family of *helix-loop-helix* (HLH) transcription factors and is responsible for the differentiation of myoblasts to myocytes, i.e., the downregulation of myoblast-specific genes and the upregulation of myocyte-specific genes.

Previous page: The picture above right shows the *Pit1* factor, a member of the *helix-turn-helix* (HTH) transcription factors family. The entire DNA-contacting surface of Pit1 is called a *POU domain*. Several other transcription factors also possess similarly large DNA-binding domains that cover a relatively long segment of dsDNA. The POU domain factors act in a similar way as factors encoded by genes called *homeobox genes*, i.e. in regulation of embryonic development.

The picture shows a piece of dsDNA (one strand in dark magenta, the other in light magenta, the phosphodeoxyribose backbone as CPKs, bases as wireframe) that is complexed over a large area with a Pit1-homodimer (one monomer in light grey, cartoons, the other in dark grey, cartoons). Of each of the two Pit1 proteins, only the dsDNA-interacting portions are shown, i.e., in each case, several helices arranged in different relative orientations to each other and contacting the dsDNA in a highly complex way.

Pit1 is responsible for the normal maturation and differentiation of the pituitary gland. A loss-of-function mutation in the pit1 gene (POU1F1) leads to congenital anterior pituitary insufficiency, i.e. HGH, prolactin, and TSH deficiency and the resultant endocrinological consequences.

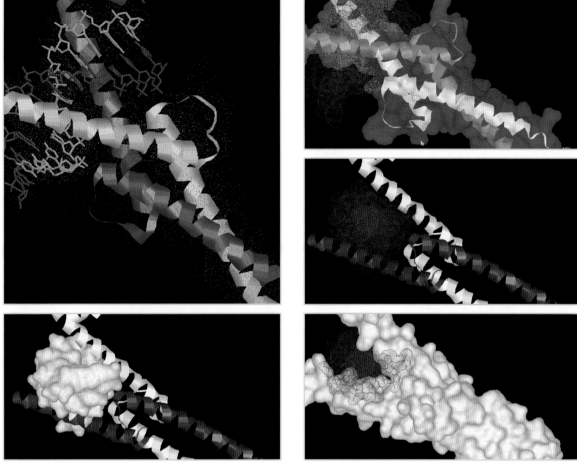

Max dimer + dsDNA (PDB 1HLO)

Max belongs to a large group of transcription factors that regulate those genes, whose products (proteins) confer upon a cell its ability to grow, i.e. to undergo cell division (mitosis, proliferation). All genes subserving this function in its broadest sense are called *oncogenes*, since any of them, when deregulated, can lead to abnormal cell proliferation, i.e., to cancer.

Like most transcription factors, Max also binds to dsDNA either as a homo- (*Max:Max*) (each monomer represented here in different greys) or as a heterodimer (e.g. *Max:Myc*). Both Max and Myc belong to the HLH superfamily, i.e., they possess helices that enable dimerization of these factors (as shown above), and further helices, suitably positioned, which permit appropriate and site-specific surrounding of a stretch of dsDNA (as also shown here). In this process, one helix positions itself into the *major groove* on one side of the dsDNA and the other helix into the next lower or upper (depending on the viewpoint) *major groove* (i.e. in the next turn), on the other side of the double helix. In this way, the dsDNA is literally wedged in and also, to some extent, bent. This bending is again the physical prerequisite for a domain of the promoter region of a gene to come into appropriate proximity to the transcription start site, i.e. where the coding region of the gene starts (that is transcribed into messenger RNA) and where the *basic transcription machinery* waits to be activated. Homodimers such as Max:Max or Myc:Myc, are often *transcription-activating* and the respective heterodimers (such as Max:Myc) *transcription-suppressing*, or the other way round, depending on the individual gene concerned, the cell type and the context of other factors present in that cell.

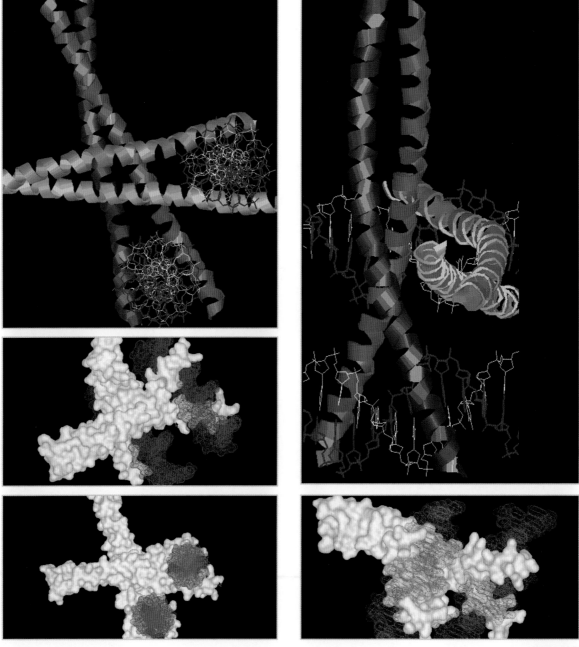

Jun-Fos heterodimer + dsDNA (PDB 1FOS)

The transcription factors *Jun* and *Fos* also belong to the class of *oncogenes*. Several *mitogenic signals* exert their effects via *phopshorylation* of Jun and Fos. Similar to Max and Myc, these two factors have specialized helices that permit dimerization and DNA binding. However, in contrast to the factors MyoD, Pit1 or Max/Myc, this helix in Jun or Fos is one and the same and therefore has to be considerably longer to spatially accommodate both functions. These two helices cross each other at a special region that is particularly rich in lipophilic leucine residues where they form a *dimerization domain* based on *hydrophobic interactions*, called a *leucine zipper*. Further away from this site, the two helices of each factor diverge to such an extent that they can wedge a stretch of dsDNA, again by immersing themselves into appropriately located *major grooves*. Jun and Fos docking sites on DNA are relatively distant from each other so that the stretch of dsDNA between them becomes, by the power exerted by the two factors, bent in a U-shape, thereby bringing both of these docking sites, together called an *AP1 site*, into a parallel orientation (as clearly seen in the picture on the right), again, a physical prerequisite for *transactivation*.

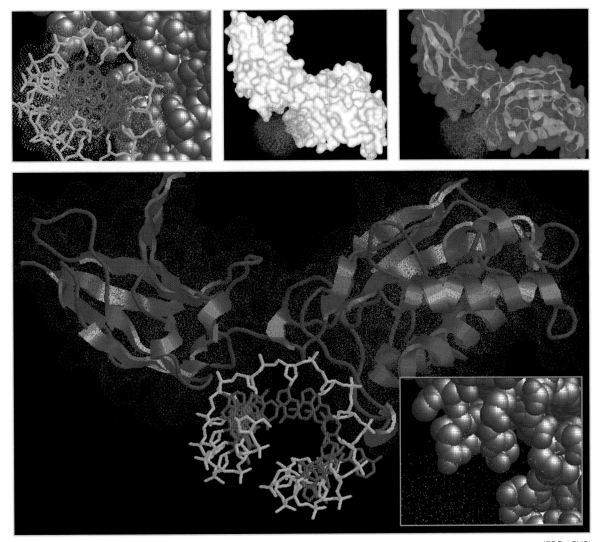

NFκB dimer ± ssDNA (PDB 1SVC)

As already stated, transcription factors are the immediate mediators of a genomic cellular response induced by external signals. Hydrophilic signals activate membrane receptors, which then interact with one or more signal transduction proteins leading to the activation of a protein kinase. This, in turn, activates a downstream transcription factor by phosphorylation, rendering it competent to interact with a suitable response element in the upstream regulatory (promoter) region of one or more genes in the affected cell, thus regulating the expression of one or, in a coordinated fashion, several genes.

For this purpose, different kinds of transcription factors employ different structural principles as to how they dimerize and how they contact dsDNA. The most stable and most suitable protein structures for the latter are α-helices, since they are highly complementary, both in macroscopic size and form, to the major groove of dsDNA.

Nuclear factor kappa B (NFκB) is one such transcription factor that plays a special role in the immune system. This picture shows how NFκB surrounds a piece of single-stranded DNA representing (half of) the NFκB-response element. In the left top panel, a segment of this interaction is zoomed in with the DNA in axial view filling the DNA-binding groove of NFκB (the empty groove is shown as inset in the right lower corner). Interestingly, the DNA-contacting domain of NFκB is constituted by several loops instead of helices.

NFκB is activated by different inflammatory signals, leading to the expression of genes, the products of which serve immune cells as receptors and, among others, histocompatibility antigens (HLA, encoded by the major histocompatibility complex, MHC). Activation of NFκb requires its dissociation from the cytosolic inhibitor protein IκB.

Transcription Factors

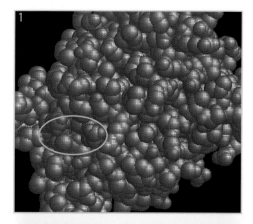
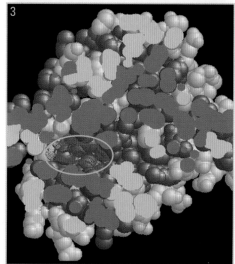
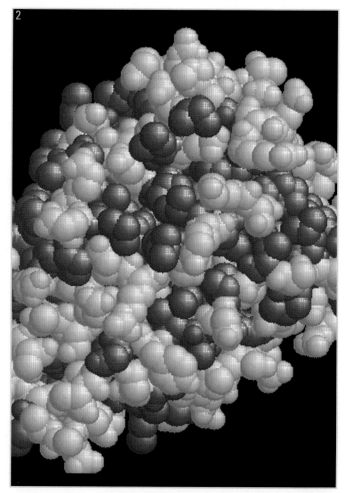

Progesterone within the **progesterone receptor** (PDB 1A28)

Steroid-dependent transcription factors (previously called *steroid receptors*) represent a further class of important regulatory proteins that occur in most, if not all, cells. They are not constitutively active, i.e. they require liganding with a steroid *agonist* (steroid *hormone*) to become transcriptionally competent. Therefore, they must possess several domains: A ligand-binding (LB) domain (LBD), a DNA-binding domain (DBD) and a transactivation domain (TAD). The pictures on this page show the LB domain of the progesterone receptor, the LB pocket filled with the ligand *progesterone*.

Panels 1 through 3 show that this portion of the protein has a globular shape consisting entirely of α-helices (panel 5 on the next page). The surface of this LB portion is predominantly hydrophilic (white atoms in panels 2, 3, 4), allowing good solubility in the cyto- and nucleosol. Very little of the ligand can be seen because the actual LB pocket is buried deep in the protein. If one were to half [by an (electronic) knife] the receptor, the pocket could be seen in its full size and internal shape, thus revealing that it is coated predominantly with lipophilic amino acids (gray atoms, panels 3, 4). This property provides an ideal environment for binding a lipophilic ligand such as a steroid hormone.

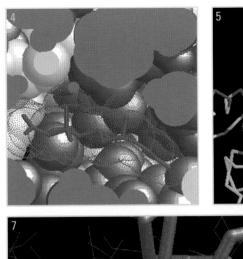
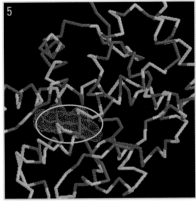
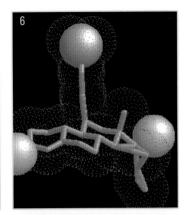
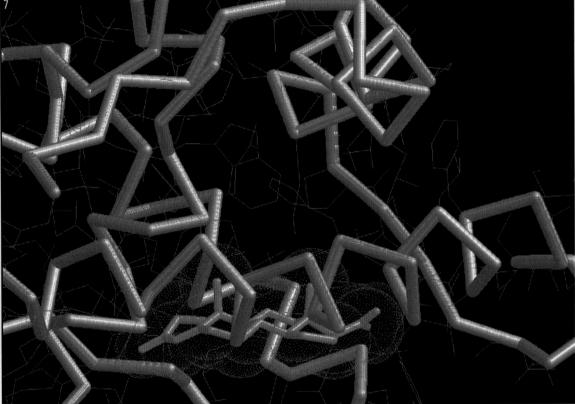

Progesterone within the **progesterone receptor**, RU486, a progesterone antagonist, shown in panel 6 (PDB 1A28)

The α-helices thus build a kind of *cage* for the ligand (panels 5 and 7). It is noteworthy that the LB pocket, with all its atoms, does not fit snugly around progesterone, but contains some unfilled space (actually filled with water molecules) precisely above the ring level of the flat-configured steroid (panels 4, 5 and 7).

If one constructs a progesterone analog that carries an additional substituent vertically to the ring level, as realized in RU486 (MIFEGYNE) (panel 6), this empty space can be filled, consequently establishing additional contacts to atoms in the pocket. This subsite within the LB pocket is called an *accessory antagonist contact site*. Additional atomic contacting (or contacts) result in a higher overall binding affinity, but concomitantly seems to hinder the necessary conformational change in the DBD and TAD of the receptor as a whole, hence appears in this case to be *unproductive*. Thus, RU486 is an antagonist and, since it binds to the same pocket as progesterone and this with even higher affinity than the agonist progesterone, it is a *competitive* antagonist.

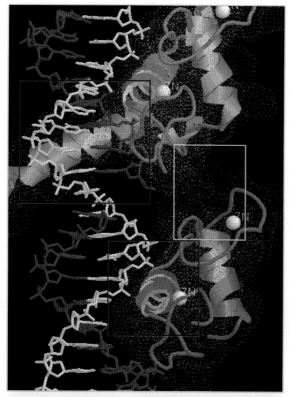

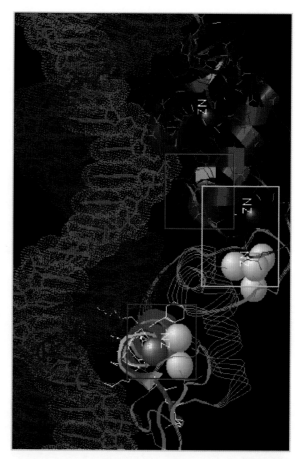

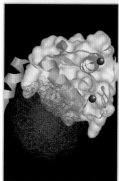 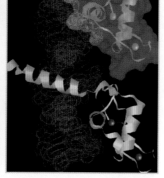

RXR + T3R + dsDNA　　　　　　　　　(PDB 2NLL)

Glucocorticoid receptor dimer + dsDNA　　　(PDB 1GLU)

These pictures show longitudinally a piece of dsDNA, of which two different neighboring sites (*response elements*) are complexed with a transcription factor. The picture on the left is the *retinoic acid X-receptor* (RXR) (above) and the *thyroid hormone receptor* (T3R) (below), whereas the picture on the right shows two identical *glucocorticoid receptors* (GR). In each case, however, only the actual DNA interacting portions of the respective proteins are shown, basically α-helices, as noted, longitudinally seen in RXR, and axially in T3R as well in both GRs. Each of these helices is positioned into a response element within a major groove of the dsDNA. Note the different orientations of the helices of the RXR-T3R heterodimer and the equal orientations in the GR homodimer (the latter two parallel to each other).

The correct relative orientations of these DNA-interacting helices within a given receptor is brought about by the zinc fingers, through which these helices are held at the main body of the protein. The zinc atoms (Zn) are shown in the left panel as white CPKs surrounded by a loop of backbone structure, and in the right panel as magenta or dark grey CPKs surrounded by light magenta sulfur atoms (only in the lower GR). Each Zn atom is coordinated in a chelate-type binding mode by the terminal γ-sulfur atoms of the four cysteine residues protruding into the interior of this loop. Furthermore, both proteins, when properly bound to dsDNA, heterodimerize or, after being properly heterodimerized, bind properly to the dsDNA. T3R can bind to DNA also as a homodimer and thus act differently. DNA-interacting domains in magenta frames, dimerization domains in white frames.

Loss of function mutations in these essential DNA-interacting amino acids lead to a loss of DNA-binding ability and thus to a loss of transcription regulating ability. Such a mutation in T3R or GR is the cause of the thyroid hormone (pseudohypothyroidism) or glucocorticoid-resistance syndromes, respectively.

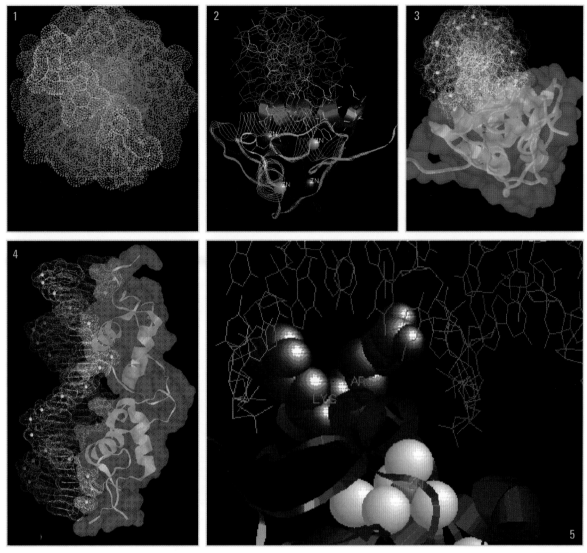

dsDNA + glucocorticoid receptor homodimer (PDB 1GLU)

These pictures show further details of the interaction between a glucocorticoid receptor (GR) and dsDNA. Panel 1 shows the axial view of dsDNA alone (the *ligand*), panels 2 and 3 show dsDNA complexed with GR dimer. It can clearly be seen that the DNA-interacting helices of both of the GRs are, vertical to the DNA's longitudinal axis, deeply immersed in the DNA's major grooves. Panel 3 shows how the GR dimer embraces the DNA half way around (other DNA proteins embrace the dsDNA even more extensively). Panel 4 shows the site in the longitudinal axis to the DNA. One can see that a certain stretch of a few nucleotides of both DNA strands each interact with a GR in a mirror-like fashion. This kind of response is called *palindromic*. Panels 4 and 5 (enlarged) show that the most intensive (or even actual) contact with DNA is established by only two AA residues out of the several present in the DNA-interacting helix. These are the basic AAs lysine and arginine lying next to each other, which, by virtue of their length, fit into and snugly fill the major groove and, by virtue of their positively charged amino groups, establish specific and atomically precise contacts to certain nucleotides on both strands, i.e. to bases via direct hydrogen bonds and/or indirect ones by intervening H_2O molecules, as well as non-specific electrostatic interactions to negatively-charged phosphate groups (for the sake of better overview, the second GR has been omitted in panel 5).

Clearly, if even a single such critical AA is mutated, the DNA-binding ability is reduced or even abrogated, resulting in a disease condition (glucocorticoid resistance).

Tumor Suppressor Proteins

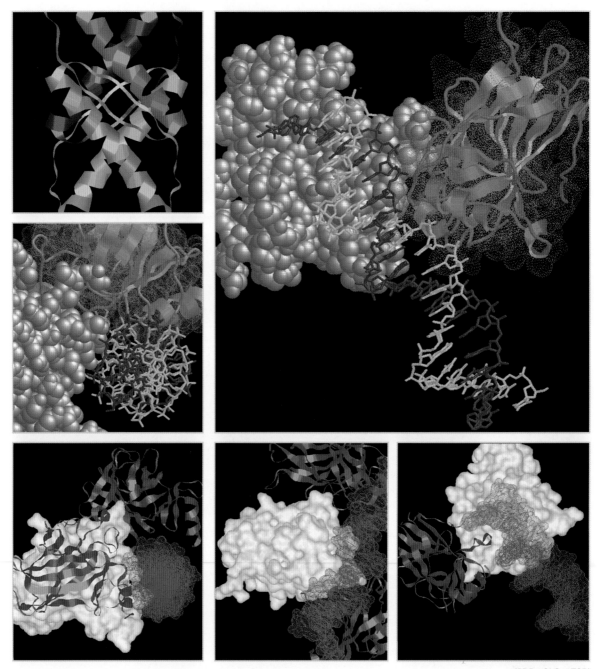

p53 core domain, homodimer, **tumor suppressor factor** in different views (PDB 1OLG, 1TSR)

Tumor suppressors are proteins that control or suppress the mitotic ability of cells. Such proteins are mostly absent in resting cells (G0-phase) or present at only very low concentrations; they need to be upregulated when a cell, triggered by a mitogenic signal, enters the mitotic cycle so that this process runs a regulated course, i.e. that after a certain number of divisions, the mitotic activity ceases rather than continues beyond a meaningful extent.

Most of the tumor suppressor factors are DNA-binding proteins and thus may be supposed to basically exert transcription-controlling functions. The pictures here illustrate this point with the p53 tumor suppressor (for better clarity, presented here as a homodimer, but in fact being a homotetramer). The picture in the left upper panel shows an interesting detail of a dimerization domain formed by α-helices that cross each other.

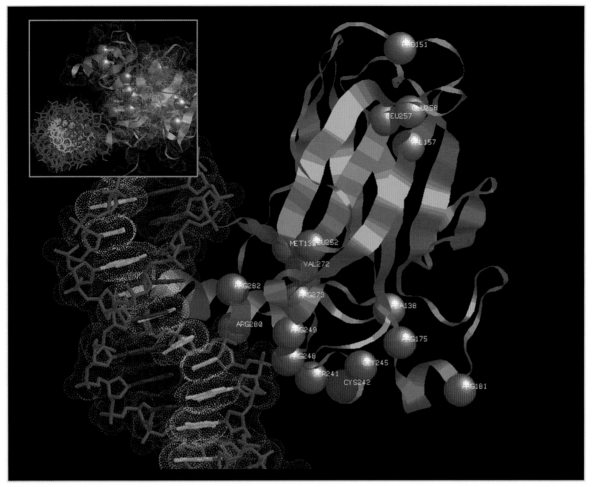

p53 dimer tumor suppressor with mutated amino acid residues highlighted (PDB 1TSR)

The p53 protein is among the most important of tumor suppressors, as can be appreciated from the fact that it is mutated in most tumors. There are somatic mutations as well as germline mutations, the latter being inherited (Li- Fraumeli syndrome).

The two pictures here (on the left, the DNA axially, and on the right longitudinally seen, complexed with one p53 monomer, for the sake of clarity) show the most frequent and significant p53 mutations [as retrieved from the OMIM (Online Mendelian Inheritance in Man) data bank]. Points of mutations are distinguished according to kind and AA sequence number. The Cα atom on a mutated AA is highlighted as CPK on the backbone. Mutations located in the DNA-interacting region of p53 are magenta and those away from that are grey.

Clearly, mutations in the DNA-contacting region of p53 statistically outnumber those elsewhere on p53. This appears quite plausible since the former mutations abrogate the DNA-interacting ability of p53, and thus apparently its mitosis-controlling and tumor-suppressing ability. However, one must keep in mind that other mutations in p53 can also be found in tumors and thus must also participate in tumor development. Obviously, such mutations, through inner conformational changes in p53, destabilize or deform the DNA interacting surfaces such that the effect would be the same as if the mutation were directly in the DNA-interacting region.

Telomere-Binding Proteins

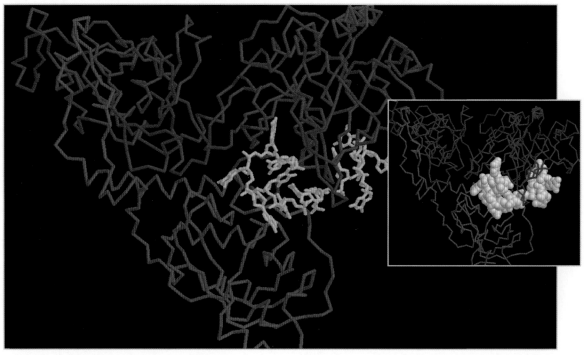

Telomere-binding protein heterodimer (PDB 1OTC)

The ends of an interphase (linear) dsDNA (as opposed to a convoluted and compacted metaphase chromosome) are free. This is typical for eukaryotic dsDNA, in contrast to prokaryotic dsDNA, which is cyclical. These ends are called telomeres. Telomeres hinder chromatid attachments to each other end-to-end as well as their susceptibility to be attacked by exonucleases. Telomeres are, therefore, essential for the chemical and structural integrity of chromatids.

Telomeres consist of specialized DNA, i.e. of tandem repeats of a consensus sequence TTTTAGGGG, and of specialized telomere-binding proteins attaching to this type of DNA, particularly at the very end. Interestingly, both the number of repeats and the composition of this sequence is species-specific. Further, it has to be considered that eukaryontic DNA polymerases can synthesize DNA only in the 3' to 5' direction. Consequently, after each DNA replication cycle, a DNA loses a short piece of telomere and it is this loss that is considered a major determinant for the maximal number of cell divisions possible in a given somatic cell of a given species. Germ cells, which should be immortal, must possess special enzymes that replace these lost pieces of DNA. These are called telomerases (actually deoxyribonucleotide transferases). In the absence of a template DNA strand, telomerases need to contain within their active center a piece of bound RNA acting as template for synthesis of DNA. This piece is 1.5 times longer than one of those repetitive DNA motifs mentioned before.

The picture above shows one telomere-binding protein consisting of an α and a β chain that together bind a piece of ssDNA (instead of RNA), i.e. 5'-GGGGTTTTGGGG-3'. Clearly, this oligonucleotide (shown in light magenta) is bent by the telomere-binding protein. Such proteins thus 'cap' the ends of a dsDNA. In addition, they appear also to display some binding affinity on their outer surface for the nuclear scaffold. The interphase chromatids thus become arranged in a certain order within the nucleus instead of 'floating' there randomly.

With advancing age, these telomere-binding proteins may become less functional (apparently by mutations within their genes) and thus cause deterioration of the functional integrity of the genome with consequent inability of cells to divide. These factors may contribute or underlie the phenomenon of cellular and body *aging*. On the other hand, activating mutations of telomerase genes may allow a cell to perform too many divisions (*cellular immortality*) and thus to undergo cancerous transformation.

Homeobox Proteins

I want to conclude this chapter on DNA-binding proteins by a recent discovery by Dr. Heiko Krude [J. Clin. Invest. 109: 475–480 (2002)], that former doctoral student in my laboratory mentioned before in the chapter on human chorionic gonadotropin receptor, who is now pediatrician and geneticist at the Charite clinic in Berlin, Germany. He, in the course of a screening programme for congenital hypothyroidism, saw five small children with clinical signs of hypothyroidism who presented, in addition and quite surprisingly, also pulmonary problems and particular neurological symptoms, i.e. choreoathetosis (involuntary and slow writhing movements). This early onset choreathetosis must have a cause different from that seen in Huntington's disease (where it has a late onset, is severe and progressive) and therefore most likely reflects an improper development of the basal ganglia of the forebrain due to a congenital defect. Since a similar benign chorea has previously been seen also in rodents with a defect in the *Nkx2-1* gene, we investigated possible defects of this gene in these patients. This gene is an evolutionary old gene, occuring already in *Drosophila,* where a loss of function mutation has been found to cause a ventral nervous system defective phenotype that – accordingly – gave the gene the name **vnd**. The vnd gene product belongs to a large family of socalled homeobox transcription factors and the gene is a socalled homeotic gene. Comparison showed that the DNA-binding homeobox domains in **vnd** and NKX2-1 are extremely similar (see sequences below), much less so being other domains of the two proteins (not shown here).

```
                    1         10        20        30        40        50
Drosophila vnd      KRKRRVLFTKAQTYELERRFRQQRYLSAPEREHLASLIRLTPTQVKIWFQNHRYKTKRAQ
human NKX2-1        R-------SQ--V-------K--K------------M-H------------------M--QA
```

Further it was shown, that NKX2-1 is identical with a protein previously named TTF1 (thyroid transcription factor 1) which was originally thought to be expressed only in the thyroid anlage and gland and now is known to be expressed also in the developing forebrain and lungs (making the name TTF1 obsolete). By sequence analysis, Heiko Krude found a heterozygous mutation in each of the five patients, i.e. one with a complete gene deletion, one with a missense mutation (G2626T) (patient A), and three with nonsense mutations, one of which had 2595insGG (patient B). The former resulted that Val45 (V) within the homeobox changed to Phe45 (F), the latter that amino acids (aa) 33 to 36 changed from A-S-M-I to G-P-A-Stop.

```
                1         10        20        30        40        50
human NKX2-1    RRKRRVLFSQAQVYELERRFKQQKYLSAPEREHLASMIHLTPTQVKIWFQNHRYKMKRQA
patient A       ------------------------------------------F-----------------
patient B       --------------------------------GPAstop
```

In oder to convert these changes in protein primary structure – as discovered by sequence analysis of the gene – into changes in the protein's native structure I could resort to the 3D structure of the homeodomain:DNA complex of vnd that was recently elucidated by NMR and deposited in the PDB (acccession code 1NK3). The model implies that the homeodomain represents a 3 helix bundle protein (aa 10–22 constituting helix 1, aa 28-38 helix 2, aa 42–60 helix 3, as indicated also by different colors in the sequence pictures above), an N-terminal random coil arm (aa 1–9) and two loops connecting helices 1 and 2 and 2 and 3, respectively, the latter representing the helix-loop-helix motiv (HLH) found in that large family of socalled HLH transcription factors examples of which examples were shown also previously in this chapter. Vnd can bind to dsDNA in a sequence-specific manner, the recognition motiv for vnd (as well as for NKX2-1) being

3'----CAAGTG-----
5'----GTTCAC-----

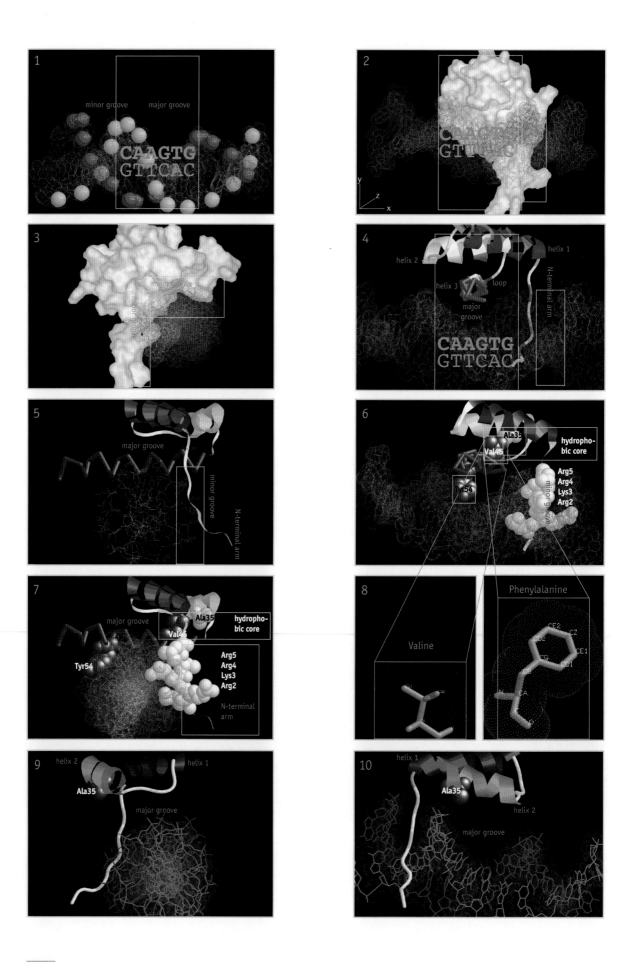

Ligand-Binding Proteins Large Molecules

Panel **1** shows a stretch of dsDNA, all atoms displayed as dot balls (CPK) except for backbone phosphor atoms (full CPK), ligh magenta one strand, dark magenta the complementary one. The grey frame overlaid to the dsDNA encompasses the sequence **CAA**GTG on one strand and GTT**CAC** on one strand and GTT**CAC** on the complementary strand.

As can be seen in panel **2**, it is this **CAACAC** motiv that is complexed with the socalled homeodomain (or "homeobox") of the transcription factor **vnd** (or NKX2-1 and related factors). Here, the homeodomain is displayed as a protein with a smoothed white surface. Note, that the largest portion of it covers the **CAACAC** motiv which itself is localized in the major groove of dsDNA. A smaller part of the homeodomain interacts with the minor groove as indicated by the ligh magenta frame.

Panel **3** shows this **vnd**:dsDNA complex turned by 90 degrees in the y-axis to the left (relative to panel 2) so as to achieve an axial view of the dsDNA.

Panel **4** shows – 45 degrees turned in the x-axis to the front relative to panel 2 – the backbone structure of **vnd** in its complex with dsDNA. It can be clearly seen that vnd is composed of 3 helices: helix 1 in dark grey, helix 2 in light grey and helix 3 in dark magenta. Note that heloix 3 represents the actual DNA-interacting helix that crosses the major groove rectangularly to the DNA longitudinal axis and fills it with its amino acid side chains. The major groove thus resembles a horse saddle in which the homedomain "sits". Panel 4 also shows that the N-terminal arm of **vnd** interacts with the minor groove coming next to the major groove **CAACAC** motiv.

Panels **5** and **7** show axial views of the dsDNA, panel **6** a longitudinal view – same orientation as in panel 2.

In panels **6** and **7**, several amino acid residues are shown as CPK: Tyr54 that represents the corner residue of the DNA-interacting helix 3 pointing to the major groove but being already outside the "saddle" (the left "leg" of the "rider"); Val45 on helix 3 that points away from the DNA and interacts with Ala35 on the oppositely located helix 2; Lys1 to Arg 5 on the N-terminal arm whose side chains partly face the minor groove and partly point away from DNA. It is clear that these positively charged amino acid side chains have a double function: those pointing to the DNA interact in part with negatively charged phosphate atoms in the backbone and thus nonspecifically stabilize the protein:DNA complex that is formed specifically by helix 3 bound to the **CAACAC** motiv, while those pointing away have been elucidated as representing the nuclear localization signal that interacts with nuclear pore proteins to allow **vnd** entry into the nucleus.

Panels **9** and **10** show two views of the modelled complex of the truncated NKX2-1 homeodomain as seen in patient B. Clearly, the entire helix 3 is missing which makes it plausible that no sequence-specific interaction is possible and therefore also no transcription regulation. What about patient A with his point mutation and change of the highly conserved Val45 to Phe45? Why is also this a loss of function mutation leading to the same phenotype as in patient B with the large truncation? As shown in panels **6** and **7**, Val45 on helix 3 points upwards towards Ala35 on helix 2 with which it forms (the greatest portion of) the protein's hydrophobic core. If Val45 is changed to a Phe, physical derangement of this core is likely. Consider that the molecular size and volume of Phe is more than twice that of Val (as shown in panel **8**), it is quite conceivable that helix 2 is lifted upwards and thereby helix 1 as well, and that thus the N-terminal arm becomes bent away from the minor groove, making the contact to the dsDNA loose, unstable or impossible at all, and thus transcritpion regulation impossible. Similar mechanisms of molecular perturbation must underly that mutant described in *Drosophila* where Ala35 is mutated to Thr35. Since Thr is hydrophilic, the hydrophobic interaction with the opppostie Val45 is severly weakened or made impossible at all. In such a case it is likely that the entire geometry of the homeodomain is disrupted, that the adoption of the native helix-bundle structure is made impossible and thus transcription regulation as well thus explaining the lethal phenotype of fruit flies afflicted with this muation [PNAS 95, 7412-6 (1998)].

Since both patients had a similar phenotype to the one with the gene deletion it can be assumed that it was in both cases the lack of 50% of NKX2-1 protein that caused the disease (a condition called haploinsufficiency). Thus, the alternative explanation, namely that the 50% of anormal NKX2-1 proteins inactivated – for example by dimerization – the 50% of normal ones (a condition called dominant negative mutation) could be ruled out.

I believe that this example perfectly illustrates the power of transferring mutational data from the gene to the native structure of the encoded protein. Understanding the general function of a given protein as well the special functions of all of its domains and subdomains will make it possible to provide a rational explanation for the type of abnormality of the gene product, i.e. loss of function vs. gain of function of the protein, respectively, and thus the mechanism leading to the particular phenotype. Such structural considerations can also explain the severity of a mutational disease. Understanding the involved protein should make it forseable which drug could be employed or first designed in order to block the unwanted effects of a mutated protein. Thus, the discovery of a new clinical syndrome (hypothyroidism, choreoathetosis, pulmonary defects) could be complemented with mechanistic and physical explanations of the molecular pathways leading to that phenotype.

Immunological Proteins

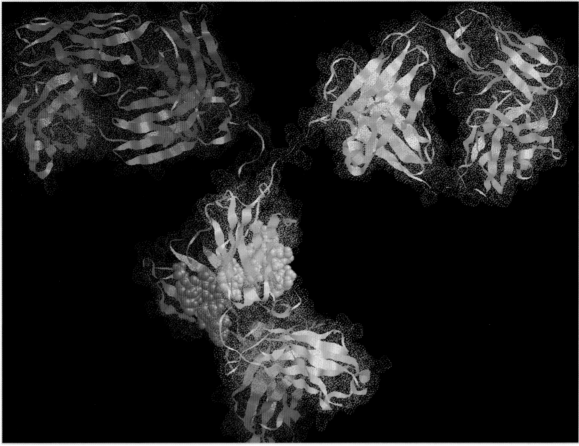

Antibody grey: light chain, magenta: heavy chains, CPKs: glycane moieties associated with the Fc truncus   (PDB 1IGT)

The raison d'être of the essential molecules of the immune system (*immunomolecules*) is the binding of *antigens*, which is the precondition for subsequent steps of the immune defense reaction in its broadest sense.

Antigens, at least potentially, are all the molecules of this universe, regardless of whether they are large or small, hydro- or lipophilic. Antigens can belong to all classes of molecular substances: mono- to polycyclic compounds, carbohydrates, proteins, nucleic acids, etc. It is thus logical that all molecules of the organism itself can be antigenic as well (endogenous or autoantigens), just as all those outside of it (exogenous antigens).

The immune system *learns*, however, to distinguish between self and nonself, so that normally it recognizes and defends the organism only against the latter. Under abnormal conditions, however, the immune system can also react with the former, leading to autoimmune diseases.

Antigens can be recognized and bound by three kinds of immunomolecules: B-cell receptors or antibodies (BCR, AB); Major histocompatibility complex (MHC) proteins, also called human leukocyte antigens (HLA), and T-cell receptors (TCR).

Further pictures, to be presented later, will show a *complement component* (C5a), a trimeric complex of the HIV-gp120-core protein bound to the T cell membrane protein CD4 as well as an antibody, and the drug *cyclosporin A* complexed with *cyclophilin*.

Antibodies and B-Cell Receptors

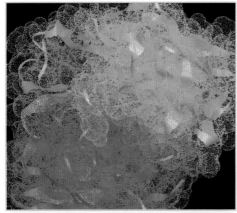

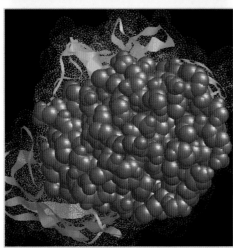

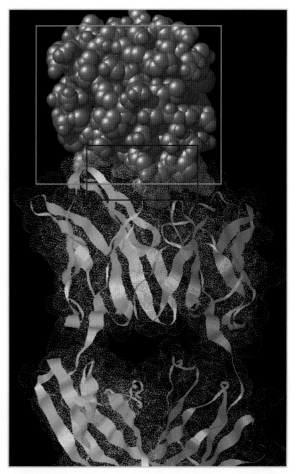

(PDB 1BQL)

Fab portion in different views (left: from above, right: axially, and from the side, longitudinally) of an **antibody** uncomplexed ('empty') and complexed with a large protein antigen, i.e. *lysozyme* (white frame, magenta frame showing the Fab-contacting *epitope* of lysozyme), grey: light chain, light magenta: heavy chain.

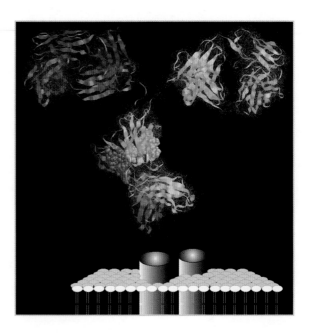

B-cell receptor: Each of the 2 heavy chains (different shades of magenta) are connected with a membrane-spanning α-helix and thus inserted into the membrane of a B lymphocyte. There, this protein serves as a receptor for binding an antigen, upon which the B cell undergoes a series of mitotic steps (clonal expansion) and then differentiates to a plasma cell that produces from the same gene an antibody. This, lacking the transmembrane modules, is consequently secreted as a soluble protein. The α-helices are removed by alternative RNA splicing. The 2 Fab portions display the same ántigen specificity as the B-cell receptor.

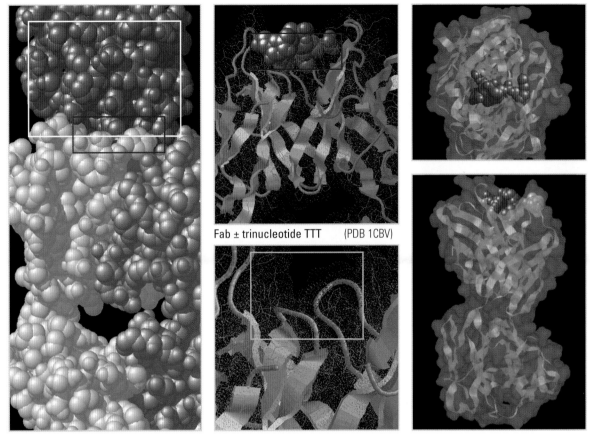

Fab ± trinucleotide TTT   (PDB 1CBV)

Fab + lysozyme   (PDB 1BQL)

Every **antibody** is Y-shaped, i.e. consists of two similar so-called *Fab arms* and one *Fc trunk*. Fab arms are comprised of two chains, a *light* and a *heavy* chain, the latter continuing to form the *Fc portion*.

The **B-cell receptor** is essentially a membrane-immobilized antibody that possesses the same antigen specificity as the respective soluble antibody. But since its heavy chains, by *alternative RNA splicing*, are elongated by a C-terminal transmembrane α-helix, the whole antibody becomes anchored in the cell membrane whereby its two Fabs and antigen-binding determinants point extracellularly. Additionally, the B-cell receptor associates laterally with special signal transduction proteins, such as CD79, consisting of an α- and a β-subunit, that, in turn, couple with their intracellular domains with protein tyrosine kinases (PTKs) as further effector proteins, e.g. *Syk* or a *Srk-type* PTK.

Principle of Endless Variability

The universality of binding possibilities of the three above-mentioned immunomolecules (with some restrictions for the MHC molecules) is based on the fact that all these molecules are *proteins* and are therefore coded by *genes*, rendering them, in a very broad sense, susceptible to *mutation*.

By a special mechanism of introducing *random gene recombinations (rearrangements)* and *point mutations* in exactly that small region of the gene of the immune molecule that codes for the antigen binding domain (the so-called *variable region*), endless variants (*iso-immunomolecules*) can be generated, each with a different antigen binding specificity. This occurs at the somatic level, i.e. in antigen-naive but otherwise immune-competent B and T lymphocytes or their precursors. Thus, an endless number of new genes develops that, however are not inheritable and thus do not *overload* the genome (30,000 genes). The major part of the immune molecule genes remains constant (*constant regions*) and confers upon these proteins a fundamentally similar molecular geometry. Pharmaceutical chemists have in recent years adopted this principle to develop new drugs for any target protein in the body possibly involved in diseases. This approach, quite opposite to *rational drug design*, is coined *random synthesis* or *combinatorial chemistry*.

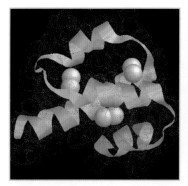

Complement component C5a (PDB 1C5A)

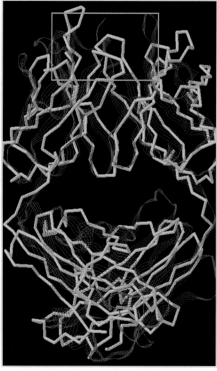

2 Fabs superimposed: one with specificity against the heptapeptide EVVPHKK (magenta, PDB 2IGF) and one against the progesterone (grey, PDN 1DBB). The picture clearly shows that the overall structure of an antibody (Fab) is constant and that only the fine structure within the antigen-binding pocket changes, depending on specificity.

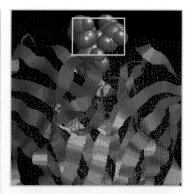
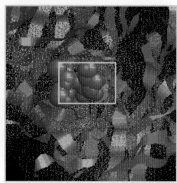

Fab + trisaccharide, (PDB 1 MFD)
seen from the side and from the top

In the antibody/B cell receptor, the variable region is formed by a β barrel, i.e. a *barrel* whose wall consists of meander-like layers of β-sheets. The antigen-binding domain represents an opening of that barrel, a circumscribed cavity in the surface of the Fab, consisting of a few AA residues. Due to combinatorial variations, and thus variable molecular geometries and physicochemical properties, these AAs account for an endless variation of *detailed solutions* in the formation of different antigen-harboring cavities and thus antigen binding specificities.

Since the overall geometry of the Fab of all antibodies is the same, i.e. the outer diameter of a Fab cylinder as well as the inner diameter of the antigen-binding pocket are fundamentally identical, antibodies can harbor in their barrels only antigens of a certain size. If the antigen is small, there is room in the barrel for the entire antigenic molecule, as can be seen in the example shown here:
- Fab + a hydrophilic trisaccharide (MW about 300),
- Fab + lipophilic progesterone (MW about 300),
- Fab + acidic hydrophilic trinucleotide (TTT, MW about 1,000, see page 89),
- Fab + basic-hydrophilic heptapeptide from myohemerythrine
  (EVVPHKK = Glu-Val-Val-Pro-His-Lys-Lys, MW about 1,000).

However, if the antigen is large (e.g. a protein), only a portion of it can be bound, i.e. only as much as can be accommodated in that Fab cavity. This *bindable portion* of the antigen is called an *epitope*. Correspondingly, the epitope-binding pocket of the antibody is called the *paratope*.

Also, the other immunomolecules (MHC, TCR) have a constant basic structure, and differing fine structures only in their respective variable regions.

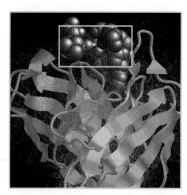

Fab + heptapeptide (PDB 2IGF)

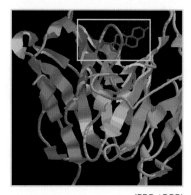

Fab + progesterone (PDB 1DBB)

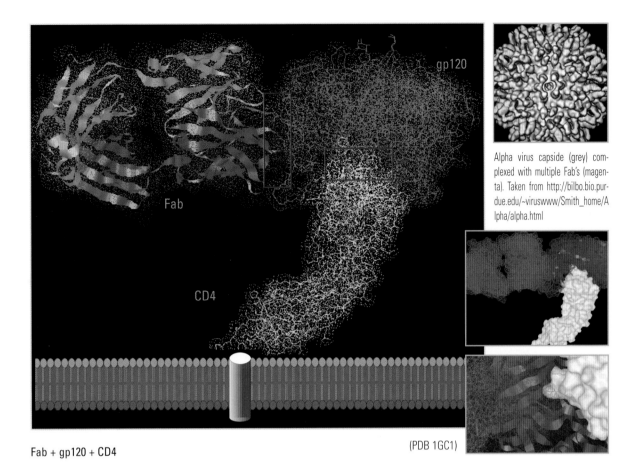

Alpha virus capside (grey) complexed with multiple Fab's (magenta). Taken from http://bilbo.bio.purdue.edu/~viruswww/Smith_home/Alpha/alpha.html

Fab + gp120 + CD4 (PDB 1GC1)

Antigen binding-induced conformational change within the Fab barrel is transmitted to more distant locations in the antibody, i.e. to its Fc portion. Thereby, binding sites become exposed (unmasked) or accessible to which proteins, for example C1, the 1st component of the *complement system*, can bind. Antibodies against cellular membrane proteins (*surface antigens*) can, provided that the latter are present in sufficient density, be brought together in such a way that C1 can be bound with high affinity between two adjacent antibodies, thus activating the entire complement cascade. This finally leads to *cell lysis* by incorporation of a *lytic complex*, which is actually a *permanently open ion channel* somewhat resembling the structure of the bee toxin *mellitin* (see chapter 'Peptidic Toxins').

Furthermore, it should be noted that the binding of an antigen or an epitope creates within the *paratope* an *induced fit*, i.e. *microconformational changes* that bring reactive AAs of the paratope closer to respective contact atoms of the antigen. This inner conformational change also extends to the *outer surface* of the Fab barrel, creating an *idiotope* or *idiotype*. As a consequence, the antibody becomes 'new' or 'foreign' to the immune system and its idiotope induces attraction of *anti-idiotypic antibodies*. The latter regulate and limit the antigen-specific immune reaction in an antigen-specific way. A large antigen can, depending on its size, possess several epitopes distributed over its entire molecular surface. Thus, a large antigen can be complexed by different antibodies, each recognizing its respective epitope. If the epitopes are sufficiently apart from each other, several antibodies can bind a single antigen even simultaneously.

Of course, each epitope on an antigen is different in shape and physicochemical properties (compare what was discussed about 'patches' on TGFβ, page xiv!) and therefore induces in each of the binding antibodies the formation of a different idiotype!

If a large protein can possess several distinct epitopes, it can just as well possess several binding domains of different kinds. As exemplified here with a portion of the *HIV-gp120* protein, one domain of gp120 binds to the CD4 protein, a membrane receptor on *T helper lymphocytes*, and another domain represents an epitope to which an antibody can bind. If the antibody comes first, it can prevent gp120 from binding to CD4. Since the HIV is infectious only through the ligand function of its gp120, such an antibody can, in principle, have an HIV-neutralizing effect and thus be protective. Please note that gp120 also contacts part of the external region of the Fab barrel. The CD4 protein is an integral membrane protein associated with the T cell receptor (TCR) and contributes to TCR signal transduction, i.e. activation of the intracellular p56 light-chain kinase.

MHC Proteins

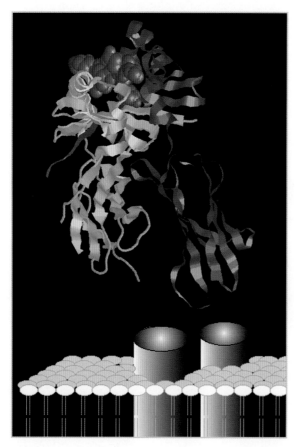 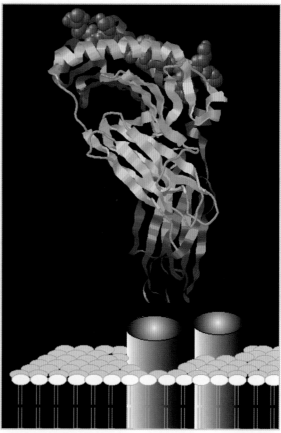

HLA-DR + presented peptide (magenta, CPKs) derived from influenza virus hemagglutinin (PDB 1DLH)

MHC class II proteins are expressed on the surface of antigen-presenting cells (APC, e.g. macrophages). One such MHC class II protein consists of two chains, α (grey) and β (light magenta), each of which traverse the cell membrane with one α-helix (here shown schematically). The extracellular domains of each of these chains interact with each other so that their distal ends form a combined superstructure representing the *bed* for a peptide to be bound and thus 'presented' on the surface.

In the genome, there are several iso-MHC II genes and in each there is also a polymorphism precisely in that bed region. Therefore, different MHC II proteins have different binding specificities for a peptide to be presented. The example shown above is a HLA-DR1 (DRB1 0101), complexed with a peptide (Pro-Lys-Tyr-Val-Lys-Gln-Asn-Thr-Leu-Lys-Leu-Ala-Thr) derived from *hemagglutinin* of an endocytosed (in contrast to a re-expressed) *influenza virus*.

APCs generally take up a complex antigen (a protein of a virus with many proteins) by receptor-mediated endocytosis and digest in endolysosomes by proteolysis the entire antigenic material into peptides of different lengths. Some of them, i.e., those with appropriate size of 10–13 AAs, are loaded onto carrier proteins and shuttled between endosomes and the plasma membrane MHC proteins and thus reach the outer face of the plasma membrane. Thus, from a single antigenic protein, several different peptides become 'presented' on the cell surface. MHC proteins are thus carriers for endogenous = cell-internal peptides, some of which are derived from cell-internal proteins and others from exogenous proteins. As MHC proteins bind peptides, they act as receptors, but since they themselves are also bound by TCRs (as shown below) they are also to be regarded as TCR ligands or agonists for the TCR.

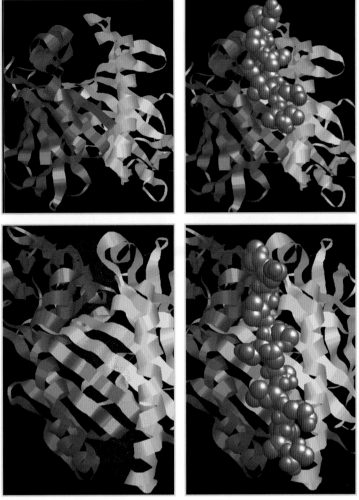

HLA-DR + presented **peptide** (magenta, CPKs) derived from influenza virus hemagglutinin (PDB 1DLH)

The pictures above show in greater detail the structure of the peptide bed within the HLA-DR1 formed, as mentioned, by parts of both chains. Together, with β-sheets, a slightly convex bottom is built, flanked on each side by a bulge of an α-helix. The linear peptide to be presented fits precisely into this bed (or cleft) along its entire length. Polymorphism (i.e. different AAs on the floor as well as in the bulges, without major architectonic differences in the global structure of this bed, is the fundamental principle of generating a huge number of different fine specificities displayed by MHC proteins for specific peptides or different ways in which a given peptide is 'bedded' and thus exposed, respectively.

MHC class I proteins are very similar to the above described MHC-II proteins. The main difference is that instead of the β-chain, a shorter second chain is complexed with the α-chain that, however, is not a transmembrane protein but a soluble accessory molecule called $β_2$-microglobulin (β2M). By its interaction, it supports the conformation of the α-chain, but does not participate in the formation of that peptide bed.

In this example, an HLA-A1 proteins filled with a peptide (Leu11-Leu-Phe-Gly-Tyr-Pro-Val-Tyr-Val19) (magenta) derived from a newly expressed TAX protein of the human lymphotropic virus type-1 (HTLV-1). As a matter of fact, the entire heterotrimer complex exists as a super-homodimer (for didactic reasons not shown).

MHC class I proteins can be found on almost all cells of the body and present peptides that are proteasome-made degradation products of endogenous cellular proteins as well as re-expressed and newly formed viral proteins. Proteasomes will be shown later, in 'Miscellaneous Macromolecules'.

T-Cell Receptor

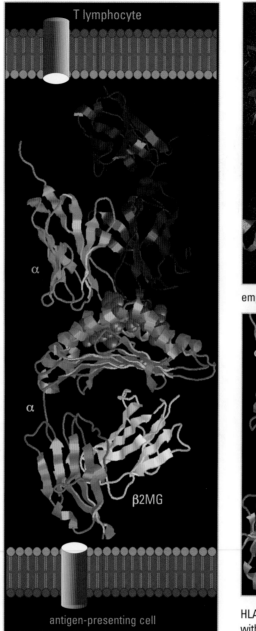
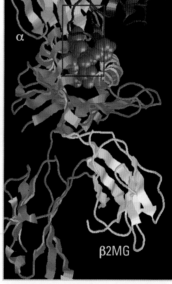

HLA-A + β2MG filled with Tax-peptide + TCR

empty HLA-A (=MHC class I) + β2MG

HLA-A + β2MG filled with Tax-peptide (PDB 1A07)

The **T-cell receptor** (TCR), as the name indicates, is expressed on the surface of T lymphocytes. The TCR itself consists of an α- and a β-chain that together, as with light and heavy chains of antibody Fab domains, form a β barrel with a distal opening, into which the antigen, more accurately the *MHC protein-presented antigen-peptide*, fits snugly (large picture to the left). Actually, the peptide through its binding to a MHC protein adopts a conformation appropriate for the TCR. Alone, it would be too flexible and thus react barely or not at all with the TCR. Conversely, the TCR recognizes the peptide only in its MHC protein-bound conformation (context).

The pictures to the right show a portion of this trimeric supercomplex, this time turned towards the front by about 30° in the y-axis (relative to the large panel on the left) without the TCR, the upper panel also without the TAX peptide (i.e. the empty MHC class I HLA-A protein).

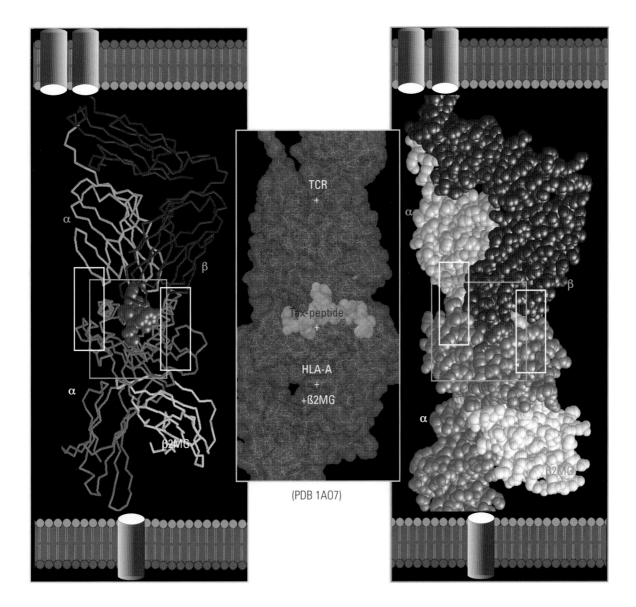

(PDB 1A07)

Furthermore, it can clearly be seen from the pictures above, especially from the one to the right, that the TCR contacts not only the presented peptide, but also the presenting MHC class I protein (α-chain). Moreover, in the TCR, both the α and β chains are engaged in contacting the MHC protein-peptide complex, whereas in the HLA-A protein, only the α-chain, not β2MG, contacts the TCR.

One can see that even a relatively small peptide can be sizable enough to function as a double ligand, i.e. on one side specifically bound by an MHC protein, and on the other simultaneously by an appropriate TCR.

So-called *superantigens* activate T cells without the requirement of being presented in the context of an MHC protein. They accomplish this either directly with the TCR without orthotopically reacting with it, or they can react non-specifically with other T-cell membrane proteins. Another mechanism is receptor *bypassing*, where they can activate intracellular mitogen pathway proteins. MMTV (*mouse mammary tumor virus*) and several bacterial toxins (*pertussis, choleratoxin*) are such superantigens.

Immunophilins

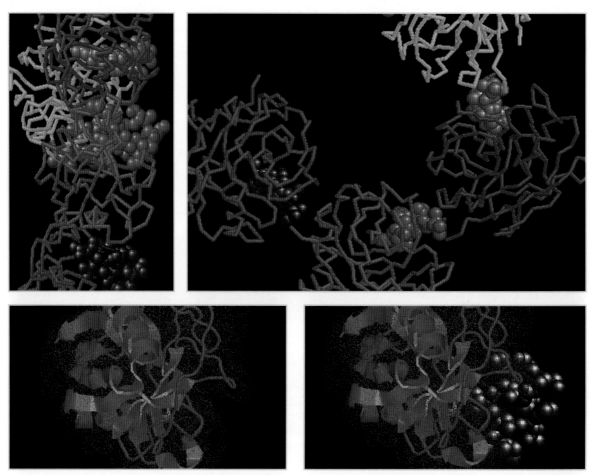

Cyclophilin tetramer ± Cyclosporin A (above) and monomer (below) (PDB 2RMA)

Cyclophilin A is a representative of the class of *immunophilins*. It is actually a *proline rotamase*, i.e. an enzyme that catalyzes the *cis/trans-isomerization* of X-proline bonds, which is one of the steps in the correct folding of a number of proteins. Cyclophilin A can be complexed by the drug *cyclosporin A* (CSA), an 11 AAs short cyclic peptide from a ground fungus (*Tolyplocadium inflatum gams*), whereupon its enzyme activity becomes inhibited. In addition, CSA can also bind to *calmodulin* and inhibit its activating function on several calmodulin-dependent proteins. A CSA-cyclophilin complex, however, can also bind to *calcineurin*, a Ca^{2+}-calmodulin-dependent *Ser/Thr protein phosphatase* expressed in neurons. Cyclosporin D can complex with unactivated steroid receptors.

Neither the *immunosuppressive mechanism* of action of CSA nor its *side effects* have yet been totally clarified. It has been proven that CSA suppresses the release of IL-1 by macrophages and also that of IL-2 by T-helper lymphocytes and thus suppresses relatively selectively the cellular branch of the immune response system. Thus, it is a specific immune suppressive drug that is now regularly administered to transplantation patients to suppress the development of a *host-graft rejection reaction*.

Miscellaneous Macromolecules
Apolipoproteins

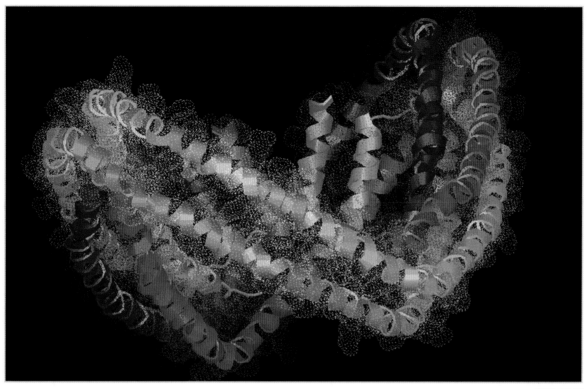

Apolipoprotein A1 (PDB 1AV1)

At the conclusion of this book, some extraordinarily shaped proteins with specially interesting structures and functions merit attention.

Apolipoproteins are the protein moieties of lipoproteins, representing a *hydrophilic shell* with a hydrophobic inner space that engulfs a superaggregate of cholesterin, fatty acid and fatty acid ester molecules. In this way, they make possible the transport of these lipophilic substances in an aqueous solution, e.g. blood. In addition, apolipoproteins serve as specific *ligands* (for an otherwise uniformly shaped mass of lipids) for specific *lipoprotein receptors*. Thus, the apolipoproteins induce receptor-mediated endocytosis of entire lipoproteins.

Mutations in apolipoproteins lower their ligand competence (affinity for the receptor) so that lipoproteins must accumulate in the blood. This leads to atherosclerosis, one of the major causes of death in developed countries of the world. Equally, there are mutations in genes for lipoprotein receptors, with similar consequences.

Apolipoproteins are also cofactors for specific enzymes, e.g. the ApoA1 shown above is a cofactor for lecithin-cholesterol-acetyltransferase. ApoA1 is the major component of the high density lipoprotein (HDL).

Mutations in the ApoA1 gene can lead to several diseases, including hypercholesterolemia. Mutations in the ApoE4 gene can lead to amyloidosis (i.e. deposition of pathological amyloid fibrils) and neurodegeneration, as seen in Alzheimer's disease.

Crystallins

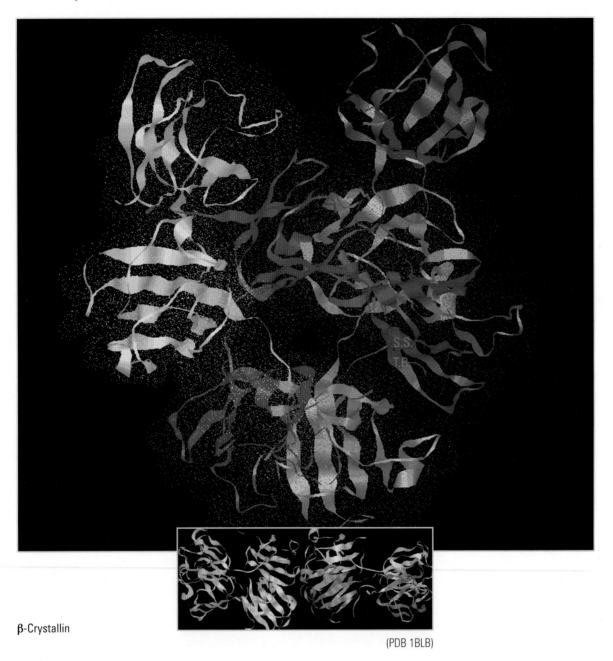

β-Crystallin

(PDB 1BLB)

Crystallins are proteins of the eye lens that have the ability to build highly complex multimers that form a gel and, above all, represent a 'crystal-clear' mass of appropriate light refraction properties. The β-crystallin shown above is a homotetramer (distinguished by different colors). Every monomer has again a 2-domain structure of β-sheets.

Some crystallins also display enzymatic activity! They are expressed in liver cells, among others, although in a 1,000-fold lower abundance than in the eye. This fact should shed light on the most interesting aspect of *gene sharing*. A single gene can, through quantitatively different degrees of expression in different cell types, produce proteins with quite different characteristics and functions, e.g. an enzyme or a structural protein.

δ-Crystallin = Arginine succinase; ε-crystallin = lactate dehydrogenase; τ-crystallin = enolase; ζ-crystallin = alcohol dehydrogenase; π-crystallin = prostaglandin synthase F, etc.

Collagens

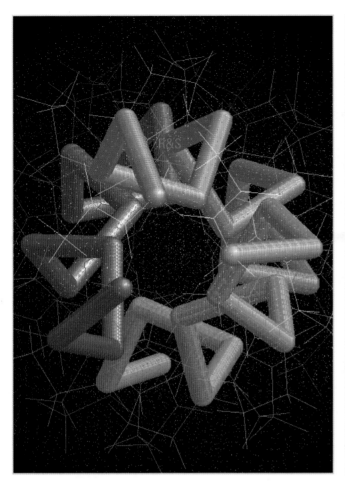
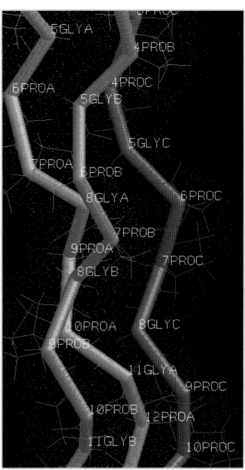

Collagen triple helix (PDB 1BBC)

Collagens represent the predominant class of extracellular structural proteins and are produced primarily by fibroblasts. There are many isotypes of collagens. Each pro-collagen peptide consists of about 1,000 amino acids, and every third one is a glycine, which makes the α-helical structure possible. Each pro-collagen possesses at the N and C termini globular moieties required for the secretability of the peptide, so that the correct *triple helix* folding can subsequently occur in the extracellular environment, after cleavage of the globular domains.

Several triple helices together form larger aggregates (*fibrils*) which, in addition, become covalently cross-linked. Fibrils with varying thickness build *fibers* with different biomechanical characteristics (elasticity, tensile strength, length) to meet different needs in different tissues. A major portion of the bone matrix consists of collagen.

Collagens are degraded by collagenases and other so-called matrix-metalloproteinases and are renewed by fibroblasts. Thus, there is a turnover and a continuous slow *remodeling* process ongoing in all connective tissue.

Mutations in the collagen genes cause severe diseases, e.g. Ehler-Danlos syndrome, epidermolysis bullosa or osteogenesis imperfecta, to mention a few.

Hemoglobin, Spectrin

Hemoglobin dimer (PDB 2HBC)

Spectrin dimer (PDB 2SPC)

The *source of life* of a multicellular organism is the possession of red blood cells (erythrocytes). These cells, through the circulatory system, carry life-giving *oxygen* from the lungs to all the organs and cells of the body, where they dispense it to them in return for the uptake of CO_2, which they transport back to the lungs for expiration.

Hemoglobin is the oxygen-binding protein that harbors a *porphyrin group* with an iron atom at its center in a specific pocket, covalently bound, as a *permanent ligand* (shown above in white CPKs). Hemoglobin occurs as a homotetramer. With increasing binding of oxygen of an individual hemoglobin monomer, the affinity for oxygen of the others in the same tetramer increases (*positive cooperativity*), similar to what was described at the beginning of this book for Ca^{2+} binding to calmodulin.

Mutations in hemoglobins are the causes of severe diseases, such as sickle cell anemia or α-thalassemia.

Spectrin is the most important cytoskeletal protein of erythrocytes. Thousands of single spectrin molecules build a spider web-like matrix below the plasma lipid bilayer membrane, which forms the erythrocytes' characteristic external cellular shape. Spectrins are α-helices whereby the individual helices cross over and bind to each other, similar to the leucine zipper motifs described previously under transcription factors.

Stress Proteins, Chaperones and Heat Shock Proteins

GroEl chaperone (PDB 1OEL)

At the end of this book, two protein multimers of highest complexity should be presented. One is a so-called *stress* or *heat shock protein*, as shown on this page. Under various kinds of stress (mechanical, physical, thermal, chemical), cells upregulate the production of certain proteins, called *stress* or *heat shock proteins* (HSPs). They all are under the control of *heat shock factors* that bind as transcription factors to *heat shock response elements* in the genes of HSPs. HSPs subserve *various protective functions*, an important feature of which appears to be the engulfment of already folded proteins to protect them from a possible heat-induced flip into an incorrect conformation, or the engulfment of nascent peptides or not yet (fully) folded or multimerized proteins following a possible incorrect folding pathway. HSP's are therefore also called *chaperon(in)es*. HSPs also serve as *protein permeases*, i.e. allow proteins to translocate through cellular membranes. HSPs are of vital importance for life and are therefore evolutionarily very old. GroEl, shown above, is present in bacteria and is an hsp60 protein. As it can be seen, GroEl is mainly built of α-helices and occurs as a homoheptamer. The entire ensemble acquires a ring- or donut-like shape with a central *channel* that harbors the peptide/protein to be protected/folded/transported. In the equatorial region, there are also holes on the side that communicate with the central channel. Chaperones are of *little binding selectivity*, i.e. they bind a vast number of different peptides/proteins. The basis for this apparent promiscuity may lie in the channel's ability to change its conformation, perhaps somewhat resembling an accordion movement, thereby providing a multitude of different interaction and binding surfaces. The only requirement for binding to GroEl appears to be the presence of an amphipathic domain, whether helical or not, on the peptide to be bound. The release of a bound peptide/protein from chaperones is generally ATP-dependent.

Proteasomal Proteins

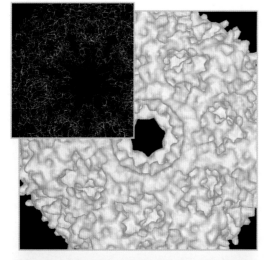

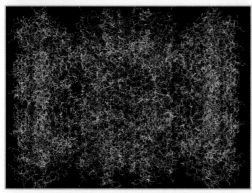

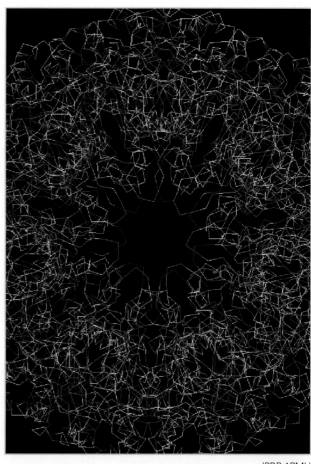

Proteasomal complex axial and longitudinal views (PDB 1PML)

Proteasomes are important protein complexes, actually multiproteases, i.e. proteolytic enzymes with multiple substrate specificities, that serve to quickly degrade within the cell all those proteins that have been improperly folded and thus may give rise to cellular damage, as well as all those that – properly folded – have to 'live' and act only shortly within a cell. To the latter belong a number of proteins that have cell cycle-regulating functions. Promitotic factors have to act only shortly, in order to guarantee well-controlled mitoses, antimitotic factors, generally of longer half life, have to be destroyed as well in order to permit cells to enter the cell cycle and finally to undergo mitosis.

In principle, all proteins that a cell produces or encounters have to be proteolytically degraded, either within the lysosomes or within the cytosol. Lysosomes also degrade proteins that a cell has taken up from the external milieu, usually by (receptor-mediated endocytosis). Proteasomes, by contrast, reside in the cytosol and degrade only endogenous proteins. Substrate proteins are distinguished, and thus proteasome-recognizable, from nonsubstrate proteins by being *ubiquitinoylated*, i.e. conjugated with one or several monomers of the small and evolutionarily extremely conserved protein *ubiq-*

uitin. Proteasomes degrade proteins to small peptides of varying lengths, some of them being appropriate in size to be translocated, by TAP proteins, into endosomes where they are loaded onto empty MHC class I proteins. The latter then shuttle to the plasma membrane and 'present' the various peptides to T cells for recognition (see chapter 'Immunological Proteins').

> Multi-ubiquitin B deposits are found in patients with Alzheimer's disease and Down's syndrome.

Proteasomes are among the largest proteins: The ubiquitin here shown is a multicatalytic protease, also named *macropain,* is composed of 28 subunits, each consisting of 233 amino acids, thus representing a molecular mass of over 600,000.

> Mutations in proteasomes can result in mitotic instability and thus cancer.

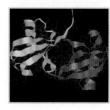

Ubiquitin-dimer   (PDB 1AAR)

Viral Coat Proteins

Foot-and-Mouth Disease Virus coat protein (PDB 1BBT)
Panels 2 and 3 were taken from http://mmtsb.scripps.edu/viper/viper.html and modified

The other supercomplex multimer protein to be shown here is a *viral capsomer protein*, e.g. that of the *foot-and-mouth disease virus* (many viruses have, in contrast to cells and bacteria, a protein capsid and not a phospholipid double layer membrane). The capsomer protein shown above is a heterotetramer (distinguished by different colors) that, as a whole, has a flat trapezoidal shape. Such trapezoids can associate side by side and and thus build a complex three-dimensionally vaulted capsid, enveloping this virus in a manner resembling the multifaceted surface of a crystal. Certain subdomains project from the surface and may serve as *ligand* structures allowing the virus to *dock* to specific *receptors on host cells* to be attacked. Other subdomains may display specific *antigenicity*, and even smaller ones may represent *epitopes* for possible *neutralizing antibodies* against this virus, as shown on page 91.

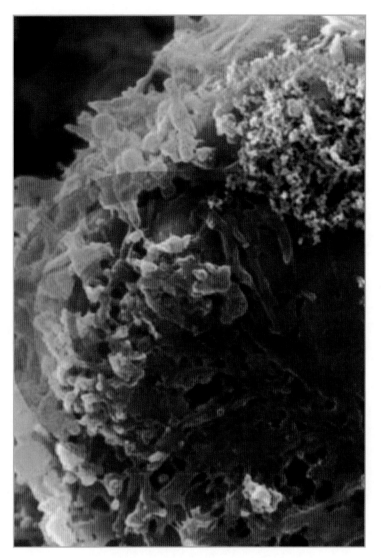

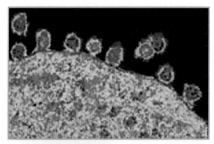

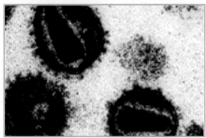

HIV daughter particles budding off from a human
T lymphocyte for a new host cell
(taken from
http://www/virology.net/Big_Virology/BVretro.
html#retro and
http://www.cmsp.com/data2/fx100003.htm)

Human T-cell leukemia virus (HTLV) infectiously attached to a T lymhocyte
(taken from Dr. Dennis Kunkel's Microscopy Science & Photography through a
Microscope; http://www.pbrc.hawaii.edu/-kunkel/gallery/viruses).

Epilogue

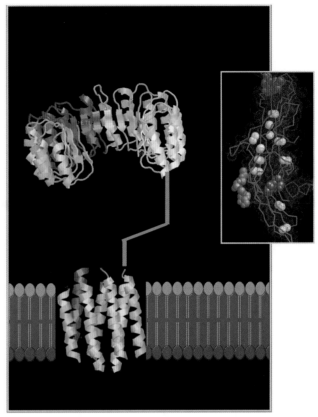

All *life* on earth is founded on *earth*, on *matter*, on *molecules*. Matter is not amorphous, but has *form* and *structure*. Structure is not randomly shaped, but has developed over millions of years of *evolution* into exquisitely interrelated forms that serve to carry out specific functions. We can see the macroscopic anatomical form of an organism, the histological make-up of an organ, the structural and ultrastructural details of a cell, and we have learned to *deduce the function from the corresponding form*. We also know that what appears large is actually composed of smaller units, e.g. a limb of an organism, a lobe in an inner organ, an organelle in a cell. These smaller units can be deformed or missing due to a disease or an accident, and yet the larger whole can be viable and functional, though with some restrictions.

The same is true for the smallest elements, the *molecules of life*, in particular the *proteins*. They consist of *domains, sites, patches, grooves, clefts*, etc. that quite frequently form *independent folding domains*. Today, we know the structure of thousands of proteins and can provide morphological descriptions and explanations. Since forms are conserved, and macroscopic forms have their corresponding microscopic and molecular forms, it is possible to *deduce function from a thoughtful observation of form*.

The pair of pictures above should illustrate this point: Through color and shape, the petals of a flower function most efficiently as traps for attracting and catching insects. We can look in the same way upon the extracellular domain of a membrane receptor as elements designed to capture a ligand. The entirety of a flower or of a receptor protein, respectively, is, to a certain extent, flexible and plastic. A conformational change induced by insects or ligands, respectively, implants itself further into the roots or into the interior of the cell, respectively.

Mutations can deform individual protein domains or be responsible for their total loss by truncation, with disease as the consequence. The *kind* and *severity* of the disease can often readily be deduced merely by knowing the domain affected and its function. The pictures of molecules in this book should enable the reader to gain an understanding of the underlying *molecular causes of diseases*.

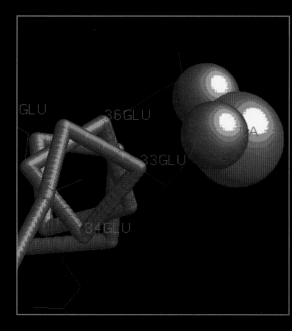

Molecules of Life

Ligands - 'Small Molecules'

1.1 Non-Peptidic 'Small Molecules'
1.2 Peptidic /Proteohormones
1.3 Peptidic Toxins

Ligand-Binding Proteins - 'Large Molecules'

2 Enzymes
3 Membrane Receptors
4 Signal Transduction Proteins
5 Multimodular Adhesion Proteins
6 DNA-Binding Proteins
7 Immunological Proteins
8 Miscellaneous Macromolecules

Understanding the structure and function of proteins should lead to understanding of the causes of *diseases*, which are basically *always molecular*, i.e. based either on *mutations* of a gene that codes for the corresponding protein, or on the *deformation/dysfunction* of a protein (encoded by a normal gene) by an *abnormal 'pathogenic' ligand*, e.g. an *autoantibody*, a *toxin* or a (toxic) *drug*, or a *destructive protease* or other influences effectuated upon that protein.

Therefore, for those with interest in medicine, the following index provides not only the page number of a given molecule described in this book, but also opens the door to several most important and authoritative data banks which will enable the reader to obtain more information and to recognize the true 'depth' of information of this book which was by didactic intention only superficially presented here. There is the the PDB accession number to the Brookhaven Protein Data Bank to obtain more details on structure, amino acid or nucleotide sequence directly at the following Internet address:

http://www.rcsb.org/pdb/cgi/queryForm.cgi.

Furthermore, the index provides the internationally-agreed gene name and its chromosomal location. With it, one can directly access the Gene Mutation Data Base of the University of Cardiff, Wales, UK:

http://archive.uwcm.ac.uk/uwcm/mg/hgmd/search.html

as well as the Gene Cards Data Bank of the Weizmann Institute in Rehovot, Israel:

http://bioinfo.weizmann.ac.il/cards/

In the former, one can obtain a list of reported gene mutations. With Gene Cards, one even opens a 'gateway' to a host of further data banks providing all the information needed.

Last but not least, and most importantly for medical doctors and students as well, the OMIM accession number is listed. With this one can enter the Online Mendelian Inheritance in Men (OMIM) Data Bank of the National Center for Biological Information in Bethesda, Maryland, USA using the following www-address:

http://www.ncbi.nlm.nih.gov/Omim/searchomim.html

from which one obtains authoritative and constantly updated information on gene structure, protein structure, allelic variants, mutations and diseases. Its value is immeasurable!

Further sources of information and internet links can be found on pages 108 and 109 of this book.

The collection of internet links given on page 108 and 109 of this book plus the entire index of this book can be found also in electronic form on the author's homepage where all the adresses and accession numbers are provided as direct hyperlinks to the respective files in the respective databanks:

http://www.uibk.ac.at/c/c5/c511/molecules_of_life_1.html
http://www.uibk.ac.at/c/c5/c511/molecules_of_life_2.html
http://www.uibk.ac.at/c/c5/c511/molecules_of_life_3.html

Updates of this book under
http://www.karger.com/molecules

Index

Ligands ('Small Molecules')

	PDB number	Gene name	Chromosomal localization	OMIM number	**Page in this book**
Nonpeptidic 'Small Molecules'					
Adenosine Phosphates (AMP, cAMP, ATP)	-	-	-	-	7
Anthracen	-	-	-	-	13
Ca²⁺ + calmodulin	1CLM	CALM1	14q24-q31	114180	3,58,59
Ca²⁺ + E-selectin	1ESL	SELE	1q23	131210	1,2,4
Calcium-ionophore: A23187	-	-	-	-	15
Mellitin (channel former)	2MLT	-	-	-	16
Catecholamines	-	-	-	-	5
Daunomycin	110D	-	-	-	14
HIV-protease inhibitor	4PHV	-	-	-	12
Mitomycin	199D	-	-	-	14
Morphine	-	-	-	-	6
Steroids (estradiol, progesterone)	-	-	-	-	8,9,76,77
Thyroid Hormones (T3)	-	-	-	-	10,11
Transthyretin, prealbumin	1THA	TTR	18q11.2	176300	11
Peptidic hormones					
Acidic fibroblast GF (aFGF), tetrameric	1AFG	FGF1	5q31	131220	22
Basic fibroblast growth factor, bFGF	1BFG	FGF2	4q25-27	134920	21
Epidermal growth factor, EGF	1EPI	EGF	4q25	131530	21
Granulocyte colony-stimulating factor, G-CSF-1	1BGD	CSF3	17q11.2	138970	20
Glucagon 1-29	1GCN	GCG	2q36	138030	18
Human chorionic gonadotropin, α:β-hCG	1HRP	CGB	19q13	118860	25,26
		CGA	6q21.1	118850	25,26
Human growth hormone	3HHR	GH1	17q22-24	139250	20
Insulin pH 7	1APH	INS	11p15.5	176730	19
Insulin, mutated	1HIQ	-	-	-	19
Insulin-like growth factor, IGF-1	1GF1	IGF1	12q22	147440	19
Interferon γ (IFNγ), dimeric	1RFB	IFNG	12q14	147570	20
Interleukin-1β (IL-1β)	2MIB	IL1B	2q14	147720	21
Nerve growth factor-β, NGFβ	1BET	NGFB	1p13.1	162030	24
Oxytocin, dimeric	1XY1	OXT	20p13	167050	18
Parathyroid hormone 1-37	1HPH	PTH	11p15.3	168450	18
Platelet-derived growth factor, PDGF A:B-dimer	1PDG	PDGFB	22q12.3	190040	24
		PDGFA	7p22	173430	24
Transforming growth factor-α, TGFα	2TGF	TGFA	2p13	190170	21
Transforming growth factor-β, TGFβ	1TFG	TGFB1	19q13.1	190180	24,i-xv
Tumor necrosis factor α (TNFα)-trimer	1TNF	TNF	6p21.3	191160	23
Tumor necrosis factor α (TNFα) + receptor	1TNR	TNFRSF1A	12p13.2	191190	23,49
Peptidic toxins					
α-Bungarotoxin + nicotinic acetylChRec-peptide	1ABT	CHRNA1	2q24	100690	28
Charybdotoxin (Scorpion)	1PTX	KCNA4	11q13.4	17626	27
Cholera toxin β subunit	1CHB	GNAS1	20q13.2	139320	29
Sarafotoxin	1SRB	EDNRA	4	131243	27
ω-Conotoxin GVIa	1OMC	GRINA	8q24.3	138251	28

Ligand-Binding-Proteins ('Large Molecules')

	PDB number	Gene name	Chromosomal localization	OMIM number	Page in this book
Enzymes					
Carbonic anhydrase + bicarbonate	1CAM	CA1	8q22	114800	31
Citric acid synthetase, dimeric	5CSC	CS	12p11	118950	33
Cyclin-dependent Kinase 2 (CdK2)	1B38	CDK2	12q13	116953	42
–/+ cyclin A	1QMZ	CCNA1		604036	42
DNA-Polymerase (KLENOW fragment) + dCTP	1KFD	POLD1	19q13.3	174761	39
DNA sliding clamp	1B77	MSH6	2p16	600678	39,67
Hexokinase ± glucose	1HKG, 2YHX	HK1	10q21	142600	33
HIV1-protease + nonpeptidic INHibitor	4PHV	-	-	-	37
HIV2-protease + renin-peptide-INHibitor	2PHV	-	-	-	36
HIV1-reverse transcriptase + nevirapin (nonpeptidic INH)	3HVT	-	-	-	38
P450 side chain cleavage enzyme + Hem + CAM	1SCC	CYP11A	15q23	118485	34
Phospholipase A2	1KVO,1BP2	PLA2G2A	1p35	172411	40
Renin + peptidic inhibitor	1SMR	REN	1q32	179820	35
RNase/angiogenin inhibitor (RAI)	1A4Y, 1BNH	RNASE6	14	601981	41
		ANG	14q11	105850	41
$3\alpha,20\beta$-HO-steroid dehydrogenase + carbenoxolone	1HDC	HSD3B1	1p13.1	109715	32
Membrane receptors, ion channels/pumps					
ATPase F1 (mitochondrial)	1BMF	-	-	-	51
hCG + putative hCG receptor	1HRP	LHCGR	2p21	152790	47,xvii-xxxii
HGH + receptor-dimer	3HHR	GHR	5p13	600946	48
		GH1	17q22	139250	20,48
IL-2 + IL-2 receptor, dimeric	1ILM	IL2	4q26	147680	48
		IL2RA	10p15	147730	48
Porin	3POR	AQP1	7p14	107776	50
Rhodopsin –/+ retinal	1BRD	RHO	3q21	180380	43-46
TNFα + receptor	1TNR	TNFRSF1A	12p13.2	191190	23,49
Signal transduction proteins					
Calmodulin	1CLM	CALM1	14q24-q31	114180	3,58
Ca-calmodulin (CAM) + CAM-binding peptide of smooth muscle MLCK	1CDL	MYLK	3cen-q21	600922	58,59
Giα (G-protein) + GTPγS,	1GFI, 1GIA	GNAI1	7q21	139310	53
Gsα (G-protein) + GDP + Mg^{2+}	1GFA	GNAS1	20q13.2	139320	55,56
Insulin receptor tyrosine kinase domain	1IRK	INSR	19q13.2	147670	57
Protein kinase A (catalytical subunit) + ATP + INH-Pep	1ATP	PRKAR2A	?	176910	57
Ras p21 + GDP + Mg^{2+}	1CRQ	HRAS	11p15.5	190020	54
Transducin α-subunit (G-protein) + GDP	1TAD	GNAT1	3p21	139330	54
SRC-homology (SH2)-domain + Phosphotyrosine	1SPS	SRC	20q12-13	190090	60
Multimodular adhesion proteins					
Cadherin module	1NCG	CDH1	16q22.1	192090	62
CD2, immunoglobulin-like Module	1CDB	CD2	1p13.1	186990	61
Fibronectin type-I module + RGD	1FBR	FN1	2q34	135600	63
Epidermal growth factor module	1APO	-	-	-	61

	PDB number	Gene name	Chromosomal localization	OMIM number	Page in this book
DNA-binding proteins					
Anti-DNA-antibody + TTT trinucleotide	-	-	-	-	70
[Daunomycin	110D	-	-	-	14]
DNA-polymerase (human) + DNA	1BPX	POLB	8p11.2	174760	68
DNA-polymerase (T7-Phage) + DNA-8mer	1T7P	-	-	-	68
Endonuclease EcoRv + dsDNA	2RVE	-	-	-	69
GATA-binding protein + GATA-Box (dsDNA)	1GAT	GATA1	Xp11.23	305371	71
Glucocorticoid receptor + dsDNA	1GLU	GRL	5q31	138040	78,79
Jun:Fos-protooncogene products + dsDNA	1FOS	FOS	14q24.3	164810	74
		JUN	1p32	165160	74
Max + dsDNA	1HLO	MAX	14q23	154950	73
[Mitomycin + dsDNA]	199D	-	-	-	67]
MyoD transcription factor + dsDNA	1MDY	MYOD1	11q15.4	159970	72
NFκB transcription factor + dsDNA	1SVC	NFKB1	4q23.	164011	75
NKX2-1 (= TTF1)	1NK3	NKX2A	14p13	600635	84
Nucleosomal complex	1AOI				67
p53 tumor suppressor factor + dsDNA	1TSR, 1OLG	TP53	17p13.1	191170	80,81
Pit-1 transcription factor + dsDNA	1AU7	POU1F1	3p11	173110	71
Progesterone receptor + progesterone	1A28	PGR	11q22	264080	76,77
RXR:T3R transcription factor, heterodimer + dsDNA	2NLL	RARA	17q12	180240	78
T3R	1BSX	THRA	3p24.3	190160	78
SRY-protein + ssDNA	1HRY	SRY	Yp11.3	480000	70
Telomere-binding protein	1OTC	TERF1	8q13	600951	82
TTT-trinucleotide + Fab (antibody)	1CBV	-	-	-	70
Immunological proteins					
Antibody complete (immunoglobulin G, IgG)	1IGT	-	*		87
B-cell receptor (immunoglobulin G, IgG)	1IGT	-	*		88
Complement-component C5a	1C5A	-	9q34.1	120900	90
Cyclophilin + cyclosporine	2RMA	PPID	4q31.3	601753	96
Fab + TTT-trinucleotide	1CBV	-	*	-	89
Fab + lysozyme	1BQL	-	*	-	88,89
Fab + progesterone	1DBB	-	*	-	90
Fab + heptapeptide	2IGF	-	*	-	90
Fab + trisaccharide	1MFD	-	*	-	90
HIV gp120-core protein + CD4 + Fab	1GC1	CD4	12pter	186940	91
HLA-DR + influenza virus-hemagglutinin peptide	1DLH	HLA-DRA	6p21.3	142860	92,93
T-cell receptor + HTLV-1-TAX peptide + HLA-A + ß2M	1AO7	TCRA	14q11.2	186880	94,95
[* Ig light chain kappa: 2q12-14, lambda: 22q12.2, heavy chain: 14q32.33]					
Miscellaneous macromolecules					
Apolipoprotein AI	1AV1	APOA1	11q23	107680	97
ß2-Crystallin, tetrameric	1BLB	CRYBA1	17q11.1	123610	98
Collagen (Gly-Pro-Pro)	1BBE	COL1A1	17q21.31	120150	99
Foot & mouth disease virus coat protein	1BBT	-	-	-	103
GroEL Chaperone	1OEL	HSPFD1	-	118190	101
Hemoglobin	2HBC	HBA1	16pter	141800	100
Proteasome 28-meric	1PMA	PSMA5	1p13	176844	102
Ubiquitin, dimeric	1AAR	UBB	17p12-p11. 1	191339	102
Spectrin modules	2SPC	SPTA1	1q11	182860	100
VIRUS PARTICLES					
Alpha virus + Fab's					91
Foot and Mouth disease virus (FMDV)					103
Human T-cell leukemia virus (HTLV)					104
Human immunodeficiency virus (HIV)					104

Internet Resources

3D Structure Data Banks and Links

PDB (Brookhaven Protein Data Bank)
http://www.rcsb.org/pdb/index.html

PDB Query Form (universal)
http://www.rcsb.org/pdb/cgi/queryForm.cgi

PDBsum, Department of Biochemistry and Molecular Biology, University College London
http://www.biochem.ucl.ac.uk/bsm/pdbsum/index.html

Research Collaboratory for Structural Bioinformatics
http://www.rcsb.org/databases.html

Molecular Biology Servers, Databases, and Web Sites of Interest
http://www.rcsb.org/pdb/web-interest.html

Education
http://www.rcsb.org/pdb/education.html

Major 3D Molecular Structure Web Sites (University of Massachusetts, Amherst)
http://www.umass.edu/microbio/rasmol/otherweb.htm

Small Molecules (Okanagan University College)
http://www.sci.ouc.bc.ca/chem/molecule/molecule.html

Klotho's Collection of Small Molecules
http://www.biocheminfo.org/klotho

PIR: Protein Information Resource
http://www-nbrf.georgetown.edu/pir/

Collections of General Molecular Biology Links

Other Online Resources for Molecular Biology
http://molbio.info.nih.gov/molbio/servers.html

Molecular Biology Data Bases
http://molbio.info.nih.gov/molbio/db.html

Pedro's BioMolecular Research Tools
http://www.public.iastate.edu/~pedro/rt_all.html

Structure of Proteins

GPCRs (G-Protein-Coupled 7 TMH-Receptors)
http://www.cmbi.kun.nl/7tm/

SWISS-PROT
http://www.expasy.ch/sidemap.html

SCOP: Structural Classification of Proteins
http://scop.mrc-lmb.cam.ac.uk/scop/index.html

NIH
Computational Molecular Biology at NIH
http://molbio.info.nih.gov/molbio/

Center for Molecular Modeling
http://cmm.info.nih.gov/modeling/

PDB At A Glance
http://cmm.info.nih.gov/modeling/pdb_at_a_glance.html

Molecules R Us
http://molbio.info.nih.gov/doc/mrus/mol_r_us.html

Software

(1) RasMol

RasMol and Chime (University of Massachusetts, Amherst)
http://www.umass.edu/microbio/rasmol/rasras.htm

RasMol-Manual
http://info.bio.cmu.edu/Courses/BiochemMols/RasFrames/TOC.HTM

(2) Other Software

Molecular Simulations Inc.
http://www.accelrys.com/products

Chemsoft
http://products.camsoft.com/reviews/

Lectures and Tutorials

Introduction to Macromolecular Simulation
http://cmm.info.nih.gov/intro_simulation/course_for_html.html

Animated Tutorials on Biomolecular Structures
http://www.umass.edu/microbio/rasmol/scripts.htm

Where to get 'Molecules' for RasMol to display?
http://www.umass.edu/microbio/rasmol/whereget.htm

RasMol Gallery
http://www.umass.edu/microbio/rasmol/galmz.htm

Rotating RasMol Images
http://www.umass.edu/microbio/rasmol/rotating.htm

Molecules Guest Book
http://molvis.sdsc.edu/guestbk/guestbk.htm

3D Molecular Tutorials
http://www.umass.edu/molvis/freichsman/index.html

Biomolecular Tutorials in Chime by Subject
http://molvis.sdsc.edu/visres/index.html

History of Visualization of Biological Macromolecules
http://www.umass.edu/microbio/rasmol/history.htm

Protein Explorer
http://molvis.sdsc.edu/protexpl/index.htm